Biocomposites - Recent Advances

Edited by Magdy M.M. Elnashar and Selcan Karakuş

Published in London, United Kingdom

Biocomposites - Recent Advances
http://dx.doi.org/10.5772/intechopen.104003
Edited by Magdy M. M. Elnashar and Selcan Karakuş

Contributors
Shahab Ahmadi Seyedkhani, Raheleh Mohammadpour, Isra Dmour, Suchita C. Warangkar, Manish R. Deshpande, Narayan D. Totewad, Archana A. Singh, Silvester Bolka, Blaž Nardin, Sedigheh Aghayari, Magdy M. M. Elnashar, Selcan Karakuş

First published in London, United Kingdom, 2023 by IntechOpen
IntechOpen is the global imprint of INTECHOPEN LIMITED, registered in England and Wales, registration number: 11086078, 5 Princes Gate Court, London, SW7 2QJ, United Kingdom

British Library Cataloguing-in-Publication Data
A catalogue record for this book is available from the British Library

Additional hard and PDF copies can be obtained from orders@intechopen.com

Biocomposites - Recent Advances
Edited by Magdy M. M. Elnashar and Selcan Karakuş
p. cm.
Print ISBN 978-1-83768-247-8
Online ISBN 978-1-83768-248-5
eBook (PDF) ISBN 978-1-83768-249-2

Meet the editors

Professor Magdy Elnashar is employed at the School of Medicine, Curtin University, Australia as well as the Centre of Excellence for Advanced Sciences, Egypt. He received his Ph.D. in Biochemistry from Leeds University, United Kingdom. Dr. Elnashar is the recipient of many awards in teaching, research, and commercialization from Australia, the United Kingdom, the United States, and Egypt. He also received the 2019 Australian Award for University Teaching. He has six patents and several research articles, books, and book chapters to his credit. His research expertise is in preparing/grafting hydrogels, immobilized enzymes, drug delivery systems, and nanoparticles/nanocomposites.

Associate Professor Selcan Karakuş is currently working in the Department of Chemistry, Istanbul University-Cerrahpasa (IUC), Turkey. She received her MSc in Physical Chemistry from Istanbul University (IU) in 2006. She received her Ph.D. in Physical Chemistry from IU in 2011. She worked as a visiting researcher in the Department of Polymer Science and Engineering, University of Massachusetts, USA. Dr. Karakuş has research experience in drug carrier systems, nanoparticles, nanocomposites, nanoemulsion self-assembled polymeric nanostructures, biosensors, and copolymer blends. She has worked on different projects funded by IUC and has published several research articles and book chapters.

Contents

Preface

This book provides a comprehensive overview of biocomposites. It discusses current research, applications, and cutting-edge technological advancements in the basic characteristics of biocomposites. It includes six chapters:

Chapter 1: "Introductory Chapter: Introduction to Biocomposites – New Insights"

Chapter 2: "Physicochemical Characterization of Nanobiocomposites"

Chapter 3: "Reactive Extrusion as an Environmentally Friendly Technology for the Production of Bio(Nano)Composites: Implementation and Characterization"

Chapter 4: "The Porosity of Nanofiber Layers"

Chapter 5: "Antibacterial, Antifungal and Antiviral Nanocomposites: Recent Advances and Mechanisms of Action"

Chapter 6: "Perspective Chapter: Tissue-Electronics Interfaces"

We truly appreciate our authors for their contributions. We would also like to express our gratitude to the editorial staff at IntechOpen for their support throughout the publication process.

Magdy M.M. Elnashar
School of Medicine,
Curtin University,
Perth, Australia

Selcan Karakuş
Faculty of Engineering,
Department of Chemistry,
Istanbul University-Cerrahpasa,
Istanbul, Turkey

Section 1

Introduction of Biocomposites and Biopolymers

Chapter 1

Introductory Chapter: Introduction to Biocomposites – New Insights

Magdy M.M. Elnashar and Selcan Karakuş

1. Introduction to Biocomposites

Biocomposites are increasingly gaining approval on the industrial scale due to their high adaptability and superior performance. Some examples of these applications are tissue engineering, drug delivery systems, restorative applications, storage devices, photocatalysts, biosensors, the encapsulation of enzymes and cells, construction, energy, rail cars, automobiles, aerospace, military applications, and packaging systems. **Figure 1** is showing some of these applications.

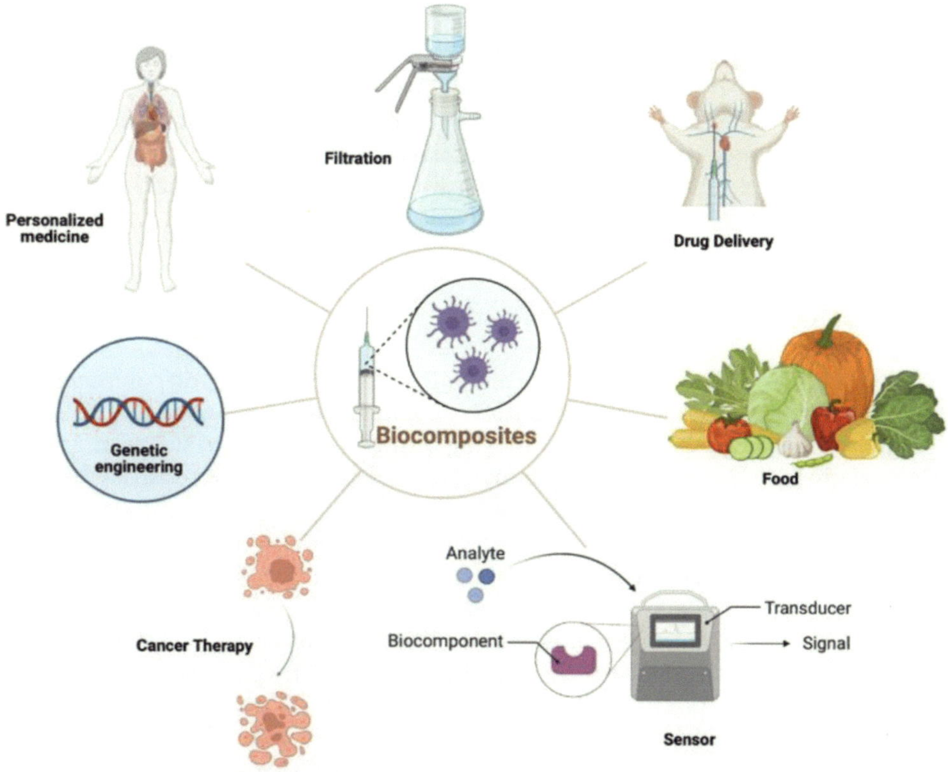

Figure 1.
Applications of biocomposites.

IntechOpen

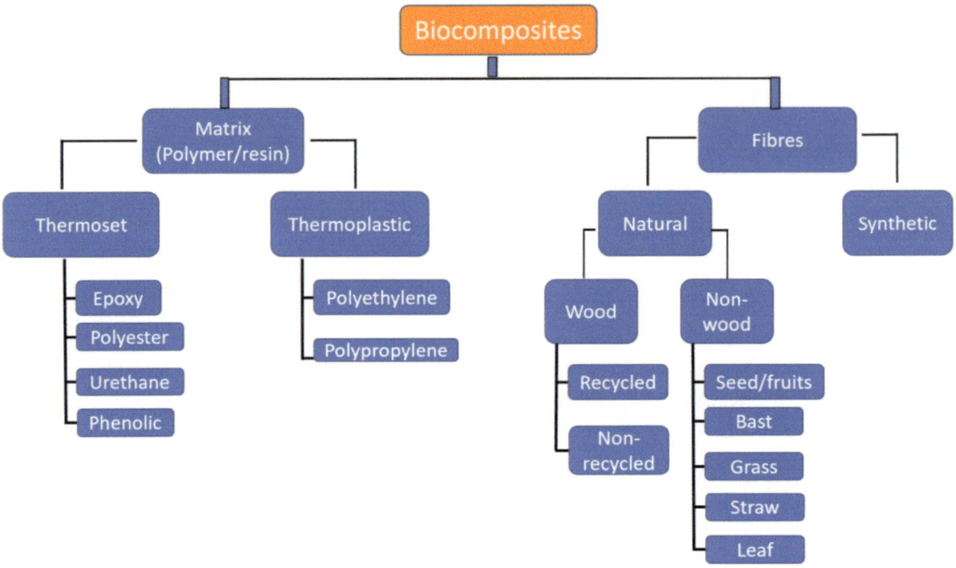

Figure 2.
Biocomposites structure.

Biocomposites are made of a matrix material (resins, biopolymers) and natural or synthetic fibers (reinforcing materials). In this context, the word "biocomposite" refers to fiber-reinforced polymer composite materials that contain bio-based fibers and/or bio-based matrix as shown in **Figure 2**. Depending on the type of polymer matrix employed, such as thermoplastics and thermosets, the characteristics of biocomposites change. The matrix plays a crucial role in holding the fibers together, transferring stresses onto them, and safeguarding them from mechanical harm and environmental deterioration.

2. Natural polymers (biopolymers)

Natural polymers (biopolymers) are more desirable than synthetic polymers due to their sustainable resources, low toxicity, biocompatibility, biodegradability, and ability to be modified, which enables tailoring of their properties to suit their application especially in the pharmaceutical industry [1]. Biopolymers are classified into three major categories as shown in **Figure 3** [2].

2.1 Applications of biopolymers

Biopolymers have a wide range of uses in the food, pharmaceutical, cosmetics, beauty, agricultural, biomedical, and many chemical industries [3–6]. The most common materials for such biopolymers are carboxymethyl cellulose (CMC), poly(amino acids), starch [7] poly(acrylamide) (PAAm), polydopamine, poly(lactide), poly(diethylaminoethyl methacrylate) (PDEAEMA), poly (acrylic acid) (PAA), poly(methacrylic acid) (PMAA), poly (ε-caprolactone) (PCL), gelatine, poly(dimethylaminoethyl methacrylate) (PDMAEMA), poly(2-methacryloyloxyethyl phosphorylcholine), albumin,

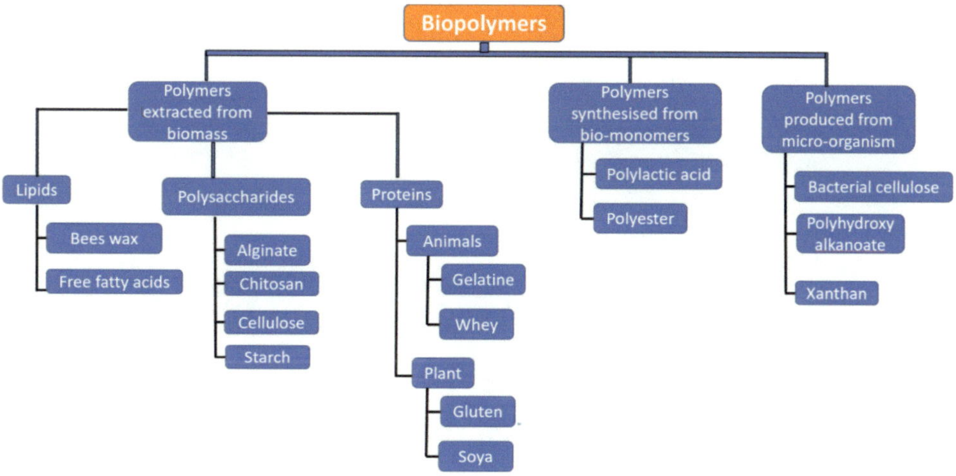

Figure 3.
Classification of biopolymers.

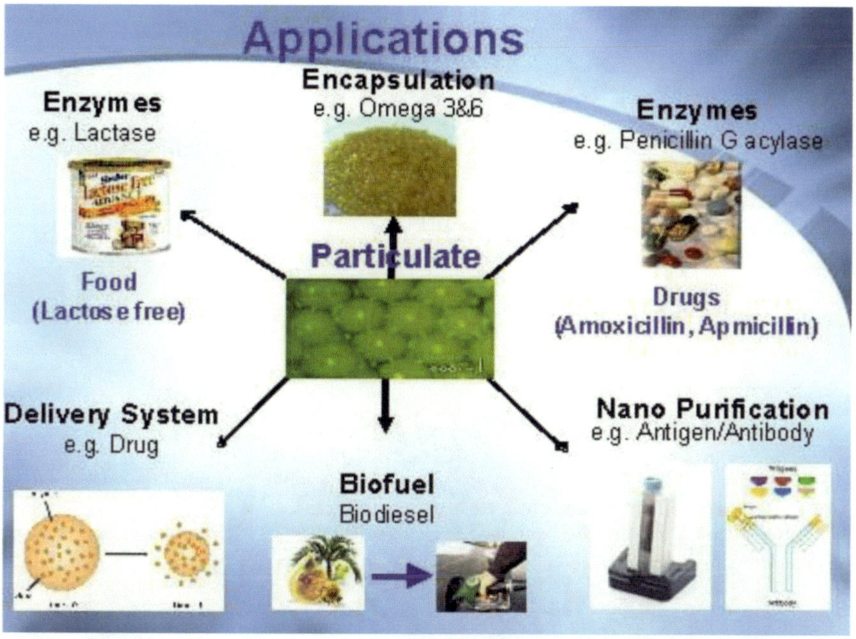

Figure 4.
Applications of grafted biopolymers.

polyvinyl alcohol (PVA), alginate, chitosan, carrageenan, and polyethylene glycol (PEG) [3, 4, 7–17]. As given in **Figure 4**, alginate [18, 19], chitosan [20–23], carrageenan [24–27] were studied, for instance, in the immobilization of enzymes (e.g. lactase), drug delivery systems, encapsulation of food (e.g. Omega 3&6), and biofuel [28, 29]. According to a forecast by European Bioplastics, the market of biopolymers will increase from 1.4 million tonnes in 2012 to roughly 6.2 million tonnes in 2027 [30].

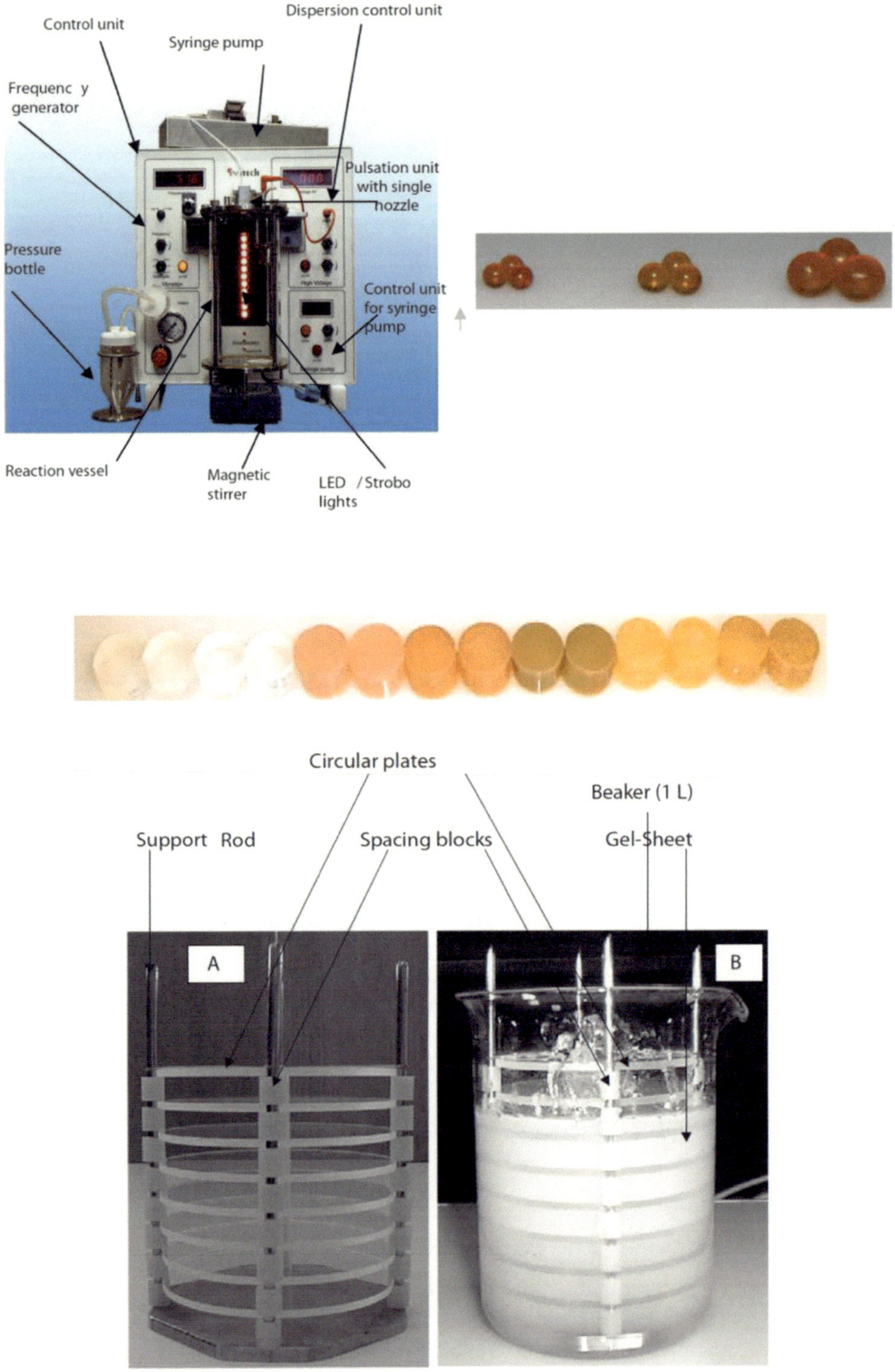

Figure 5.
Particulates prepared in the size of macro to nano including magnetized beads. Pictures reproduced from Elnashar (a) production of micro beads using the encapsulator [3], and (b) production of uniform gel sheets and disks using the parallel plate equipment [31, 32].

2.2 Shapes of biopolymers

The biopolymers can be treated and produced in different shapes using the Innotech Encapsulator, vibrational jet-flow technology, ionic-gelation methods, parallel plates, dripping and interphase techniques. Gel sheets, disks, and beads were produced using the Innotech Encapsulator and the parallel plates as shown in **Figure 5** [3, 31]. Nanoparticles can be synthesized using a variety of techniques that fall under the top-down or bottom-up method categories [33].

Nanotechnology has become an attractive research field with a high potential in field of development of advanced nano-products due to their superior surface, physical, chemical, biological, and mechanical properties. That was also due to their nano-size, morphology, shape, solubility, biodegradability, and biocompatibility. Particularly, bio/nanocomposites have high surface to volume ration (small particles ranging from 1 to 100 nm in size), thus the interaction between the matrix (shell/biopolymer) and reinforcement (core/fiber) is particularly strong [34].

3. Fibers

The fibers provide strength and stiffness to the structure. The classification of fibers is presented in **Table 1**. Fibers are obtained either naturally or man-made as shown in **Table 1**. Although synthetic and natural fibers can be used in biocomposites, the use of natural fiber as reinforcement in polymeric composites has been preferred due to environmental concerns and the high cost of synthetic fibers [35, 36]. Naturally occurring fibers can be classified into three main categories: mineral, animal, and plant fibers. The latter are the most abundant fibers among all the natural fibers. Silk fiber (animal fiber) has the highest tensile strength among all the natural fibers [37]. Whereas asbestos and ceramic (mineral fibers) can function in high temperatures [38]. More than 65% of natural fiber-based composites are used in the packaging sector, with the remaining 35% being used in the medical, textile, electrical, and agricultural sectors [39].

Biocomposites							
	Wood fibers		**Non-wood natural fibers**				
	Recycled	**Non-recycled**	**Seed/ fruits**	**Bast**	**Grass**	**Straw**	**Leaf**
Examples	Papers, magazines, newspapers fibers	Soft and hard wood	Cotton, coconut, coir	Help, flax, jute, kenaf	Switch grass, elephant grass, bamboo, bamboo fiber	Wheat, corn, rice, straw	Pineapple leaf, sisal, henequen

Table 1.
Classification of biocomposites' fibers.

4. Future overview

Recent developments in the study of bio-based nanostructures, with an emphasis on the environmentally sustainable production of biocomposites, have tremendously benefited biomedical applications. The low-cost, environmentally beneficial method of creating biocomposites with biodegradable and biocompatible polymers currently has a variety of advantages. The advantages of pH and temperature sensitive dual-stimuli responsive nano-systems should be investigated by next-generation drug delivery systems using in vitro or in vivo methods. These discoveries imply that additional research in this field is required given the lack of experimental knowledge and understanding of drug release systems, bio-sensing mechanism, and therapeutic effects of these biocomposites.

Author details

Magdy M.M. Elnashar[1,2]* and Selcan Karakuş[3]

1 School of Medicine, Curtin University, Perth, Australia

2 Centre of Scientific Excellence—Polymers, Department—Advanced Materials and Nanotechnology Laboratory, National Research Centre, Cairo, Egypt

3 Faculty of Engineering, Department of Chemistry, Istanbul University-Cerrahpasa, Istanbul, Turkey

*Address all correspondence to: magmel@gmail.com; magdyelnashar@curtin.edu.au

References

[1] Dmour I, Taha MO. Natural and semisynthetic polymers in pharmaceutical nanotechnology. Organic Materials as Smart Nanocarriers for Drug Delivery. 2018;**2018**:35-100. DOI: 10.1016/B978-0-12-813663-8.00002-6

[2] Lalit R, Mayank P, Ankur K. Natural fibers and biopolymers characterization: A future potential composite material. Strojnicky Casopis. 2018;**68**:33-50. DOI: 10.2478/SCJME-2018-0004

[3] Elnashar M, editor. Biopolymers. 28 Sep 2010. Available from: http://dx.doi.org/10.5772/286

[4] Elnashar M, editor. Biotechnology of Biopolymers. 5 Jul 2011. Available from: http://dx.doi.org/10.5772/683

[5] Elnashar MMM. The art of immobilization using biopolymers, biomaterials and nanobiotechnology. Biotechnology of Biopolymers. InTech; 2011. DOI: 10.5772/23696

[6] Awad GEA, Amer H, El-Gammal EW, et al. Production optimization of invertase by lactobacillus brevis Mm-6 and its immobilization on alginate beads. Carbohydrate Polymers. 2013;**93**:740-746. DOI: 10.1016/J.CARBPOL.2012.12.039

[7] Khalili H, Bahloul A, Ablouh EH, et al. Starch biocomposites based on cellulose microfibers and nanocrystals extracted from alfa fibers (Stipa tenacissima). International Journal of Biological Macromolecules. 2023;**226**:345-356. DOI: 10.1016/J.IJBIOMAC.2022.11.313

[8] Salimi M, Motamedi E, Motesharezedeh B, et al. Starch-g-poly(acrylic acid-co-acrylamide) composites reinforced with natural char nanoparticles toward environmentally benign slow-release urea fertilizers. Journal of Environmental Chemical Engineering. 2020;**8**:103765. DOI: 10.1016/J.JECE.2020.103765

[9] Guo D. Effect of electron beam radiation processing on mechanical and thermal properties of fully biodegradable crops straw/poly (vinyl alcohol) biocomposites. Radiation Physics and Chemistry. 2017;**130**:202-207. DOI: 10.1016/J.RADPHYSCHEM.2016.08.024

[10] Hu H, Zhang Z, Fang Y, et al. Therapeutic poly(amino acid)s as drug carriers for cancer therapy. Chinese Chemical Letters. 2022;**2022**:107953. DOI: 10.1016/J.CCLET.2022.107953

[11] Lin Z, Fu H, Zhang Y, et al. Enhanced antibacterial effect and biodegradation of coating via dual-in-situ growth based on carboxymethyl cellulose. Carbohydrate Polymers. 2023;**302**:120433. DOI: 10.1016/J.CARBPOL.2022.120433

[12] Xiong S, Wang Y, Zhu J, et al. Poly (ε-caprolactone)-grafted polydopamine particles for biocomposites with near-infrared light triggered self-healing ability. Polymer (Guildf). 2016;**84**:328-335. DOI: 10.1016/J.POLYMER.2016.01.005

[13] Gawad RMA, Kattab HM, Strabel M, et al. Effect of different levels from linseed oil and linseed oil beads on rumen fermentation and microbial parameters using gas production system and rumen simulation technique. Asian Journal of Animal and Veterinary. 2015;**10**:97-118. DOI: 10.3923/AJAVA.2015.97.118

[14] Awad GEA, Abd El Aty AA, Shehata AN, et al. Covalent immobilization of microbial

naringinase using novel thermally stable biopolymer for hydrolysis of naringin. Biotech. 2016;**6**:1-10. DOI: 10.1007/S13205-015-0338-X/FIGURES/8

[15] Elnashar MM, Mohamed EH. Novel epoxy activated hydrogels for solving lactose intolerance. BioMed Research International. 2014;**9**:1-9. DOI: 10.1155/2014/817985

[16] Ghosh S, Abanteriba S, Wong S, Houshyar S. Performance analysis of grafted poly (2-methacryloyloxyethyl phosphorylcholine) on additively manufactured titanium substrate for hip implant applications. Journal of the Mechanical Behavior of Biomedical Materials. 2019;**100**:103412. DOI: 10.1016/J.JMBBM.2019.103412

[17] Kanth S, Malgar Puttaiahgowda Y, Nagaraja A, Bukva M. Recent advances in development of poly (dimethylaminoethyl methacrylate) antimicrobial polymers. European Polymer Journal. 2022;**163**:110930. DOI: 10.1016/J.EURPOLYMJ.2021.110930

[18] Gorshkova MY, Vanchugova L, Volkova IF, et al. Novel mucoadhesive carriers based on alginate-acrylamide hydrogels for drug delivery. Mendeleev Communications. 2022;**32**:189-191. DOI: 10.1016/J.MENCOM.2022.03.012

[19] Hamed SF, Hashim AF, Abdel Hamid HA, et al. Edible alginate/chitosan-based nanocomposite microspheres as delivery vehicles of omega-3 rich oils. Carbohydrate Polymers. 2020;**239**:116201. DOI: 10.1016/J.CARBPOL.2020.116201

[20] Fard GH, Moinipoor Z, Anastasova-Ivanova S, et al. Development of chitosan, pullulan, and alginate based drug-loaded nano-emulsions as a potential malignant melanoma delivery platform. Carbohydrate Polymer Technologies and Applications. 2022;**4**:100250. DOI: 10.1016/J.CARPTA.2022.100250

[21] Ribeiro ES, de Farias BS, Santanna C, Junior TR, et al. Chitosan–based nanofibers for enzyme immobilization. International Journal of Biological Macromolecules. 2021;**183**:1959-1970. DOI: 10.1016/J.IJBIOMAC.2021.05.214

[22] El-Kady AM, Kamel NA, Elnashar MM, Farag MM. Production of bioactive glass/chitosan scaffolds by freeze-gelation for optimized vancomycin delivery: Effectiveness of glass presence on controlling the drug release kinetics. Journal of Drug Delivery Science Technology. 2021;**66**:102779. DOI: 10.1016/J.JDDST.2021.102779

[23] Latif AAN. Chitosan-benzofuran adduct for potential biomedical applications: Improved antibacterial and antifungal properties. Scholars Research Library Der Pharmacia Lettre. 2015;**7**:107-117

[24] Geyik G, Işıklan N. Design and fabrication of hybrid triple-responsive κ-carrageenan-based nanospheres for controlled drug delivery. International Journal of Biological Macromolecules. 2021;**192**:701-715. DOI: 10.1016/J.IJBIOMAC.2021.10.007

[25] Hambleton A, Fabra MJ, Debeaufort F, et al. Interface and aroma barrier properties of iota-carrageenan emulsion–based films used for encapsulation of active food compounds. Journal of Food Engineering. 2009;**93**:80-88. DOI: 10.1016/J.JFOODENG.2009.01.001

[26] Elnashar MMM, Yassin MA. Covalent immobilization of β-galactosidase on carrageenan coated with chitosan.

Journal of Applied Polymer Science. 2009;**114**:17-24. DOI: 10.1002/APP.30535

[27] Ali KA, Hassan ME, Elnashar MMM. Development of functionalized carrageenan, chitosan and alginate as polymeric chelating ligands for water softening. International journal of Environmental Science and Technology. 2017;**14**:2009-2014. DOI: 10.1007/S13762-017-1298-Y/METRICS

[28] Elnashar MMM, Yassin MA. Lactose hydrolysis by β-galactosidase covalently immobilized to thermally stable biopolymers. Applied Biochemistry and Biotechnology. 2009;**159**:426-437. DOI: 10.1007/S12010-008-8453-3/METRICS

[29] Elnashar MM, Awad GE, Hassan ME, et al. Optimal immobilization of β - galactosidase onto -carrageenan gel beads using response surface methodology and its applications. The Scientific World Journal. 2014:1-7. DOI: 10.1155/2014/571682

[30] Global bioplastics production will more than triple within the next five years – European Bioplastics e.V. Available from: https://www.european-bioplastics.org/global-bioplastics-production-will-more-than-triple-within-the-next-five-years/ [Accessed December 21, 2022]

[31] Elnashar MMM, Hassan ME, Awad GEA. Grafted carrageenan gel disks and beads with Polyethylenimine and glutaraldehyde for covalent immobilization of penicillin G Acylase. Journal of Colloid Science and Biotechnology. 2013;**2**:27-33. DOI: 10.1166/JCSB.2013.1029

[32] Elnashar MM, Millner PA, Johnson AF, Gibson TD. Parallel plate equipment for preparation of uniform gel sheets. Biotechnology Letters. 2005;**27**:737-739. DOI: 10.1007/S10529-005-5363-0/METRICS

[33] Wagner Ferreira Sabará E, Pereira V, Luiz Molisani A, et al. A review on the classification, characterisation, synthesis of nanoparticles and their application. IOP Conference Series Materials Science Engineering. 2017;**263**:032019. DOI: 10.1088/1757-899X/263/3/032019

[34] Mahmood T, Ullah A, Ali R, et al. Improved nanocomposite materials and their applications. Nanocomposite Materials for Biomedical and Energy Storage Applications. 2022:1-23. DOI: 10.5772/INTECHOPEN.102538

[35] Sreenivasan VS, Rajini N, Alavudeen A, Arumugaprabu V. Dynamic mechanical and thermo-gravimetric analysis of Sansevieria cylindrica/polyester composite: Effect of fiber length, fiber loading and chemical treatment. Composites Part B Complete. 2015;**2015**:76-86. DOI: 10.1016/J.COMPOSITESB.2014.09.025

[36] Bakar NA, Chee CY, Abdullah LC, et al. Thermal and dynamic mechanical properties of grafted kenaf filled poly (vinyl chloride)/ethylene vinyl acetate composites. Materials & Design. 2015;**2015**(65):204-211. DOI: 10.1016/J.MATDES.2014.09.027

[37] Rojo E, Alonso MV, Oliet M, et al. Effect of fiber loading on the properties of treated cellulose fiber-reinforced phenolic composites. Composite B Engineering. 2015;**68**:185-192. DOI: 10.1016/J.COMPOSITESB.2014.08.047

[38] Cai M, Takagi H, Nakagaito AN, et al. Effect of alkali treatment on interfacial bonding in abaca fiber-reinforced

composites. Composites. Part A, Applied Science and Manufacturing. 2016;**90**:589-597. DOI: 10.1016/J.COMPOSITESA.2016.08.025

[39] Biopolymer production for (petro-) chemical sector, Climate Technology Centre & Network. 2016. Available from: https://www.ctc-n.org/technologies/biopolymer-production-petro-chemical-sector [Accessed December 21, 2022]

Section 2

Synthesis and Characterization of Biocomposites

Chapter 2

Physicochemical Characterization of Nanobiocomposites

Isra Dmour

Abstract

Nanobiocomposites (NBCs) have many applications in drug delivery, tissue engineering, etc. The need for NBC physicochemical characterization is mandatory before investigating their usefulness in developing drug delivery systems. This chapter will explore the basic and the most recent techniques used in the physicochemical characterization of these biocomposites. Examples of physical properties include morphological properties using microscopy (size, porosity, etc.), particle size analysis and surface charge, powder X-ray diffraction, thermal, mechanical, and rheological properties, etc. Examples of chemical properties include molecular weight determination, solubility and purity assessment, degree of functionalization, and gelling properties, using spectroscopic techniques (UV, MS, NMR, etc.). For each property, the following points will be elucidated: sample preparation, factors affecting the accuracy of the test results, examples of data interpretation from the recently published literature, and test limitations, if any.

Keywords: nanobiocomposite, biocomposite, physicochemical, characterization, drug delivery, nanoparticle

1. Introduction

Biodegradable polymers offer great potential in drug delivery using nano-scale systems. However, natural polymers are more attractive for pharmaceutical applications as they have sustainable resources, low toxicity, biocompatibility, biodegradability, and the ability to be modified, allowing tuning of their properties to suit their application in the pharmaceutical field [1]. Many polysaccharide polymers have been investigated for drug delivery application, including chitin/chitosan, agarose, Bacterial cellulose, gum arabic, tragacanth, alginate, gellan gum, starch, carrageenan, dextran, nanocellulose, and Xanthan gum [2].

On the other hand, biocomposite materials consist of at least two components, including a continuous matrix phase, which is usually a natural polymer, and discontinuous reinforcement material, which will be used to reinforce the backbone of the biopolymer. Polymers with counter ionic properties like cellulose and chitosan or using a crosslinker like tripolyphosphate (TPP) are commonly used [3]. Reinforcement can be performed using physical crosslinking methods, which typically include electrostatic/ionic interactions, hydrophobic interactions, and π–π stacking interactions. In comparison, chemical crosslinking methods are typically covalent crosslinking, which h can be a direct crosslinking or free-radical polymerization [4, 5]. For example, the fabrication of chitosan NPs using tripolyphosphate crosslinking, followed by covalent crosslinking using coupling chemistry [6, 7]. As a result, the size and properties of the NBCs offer excellent features in the drug delivery field [8]. In addition, micro to nano particles, can be prepared by a variety of technologies, including the Innotech Encapsulator, ionic-gelation techniques, vibrational jet-flow technology, dripping, and interphase technique approach [9], can be used to create the tiny micro- and nanocapsules [10, 11]. On the other hands, nanogels can be prepared in uniform gel sheets or in macro disks using the parallel plates equipment [12].

In nanocomposites, the interaction between matrix and reinforcement is very high due to the high surface-to-volume ratio [8, 13]. The improved properties of nanocomposites depend on the properties of each component, their relative amounts, and the overall geometry of the nanocomposites. Generally, when natural polymers are utilized in the fabrication of NBCs, enhanced properties, such as NP size and surface charge, mucoadhesiveness, adsorptive, etc., can be achieved [1, 3]. These characteristics can be optimized to suit a wide range of applications in non-conventional routes of administration, including nasal, rectal, buccal, etc. In addition, many of these NBCs have been fabricated as stimuli-responsive drug delivery systems (pH, temperature, light, etc.) [14, 15].

This chapter focuses on the physicochemical characterization of polysaccharides-based biocomposites investigated as nano drug delivery systems. Before utilizing a new polymer in NBCs fabrication, a complete characterization should be performed using reliable and validated methods. Then, biocomposites can be characterized using traditional and advanced methods, including spectroscopic techniques like Infrared (IR), Nuclear Magnetic Resonance (NMR), etc. Additional properties like thermal, mechanical, rheological, gelling, adsorptive, etc. are also part of the characterization procedure commonly reported. Representative examples from the most recently published literature will also be discussed. **Figure 1** proposes a plan to follow in the physicochemical characterization of natural polymers and their corresponding NBCs.

2. Chemical characterization

2.1 Ultraviolet: Visible (UV: VIS) spectroscopy

UV–VIS analysis is a simple, low-cost, and rapid technique that is based on exposing the biocomposite sample to electromagnetic radiation in the UV–Vis region (typically 190 to 900 nm). The UV range usually extends from 100 to 400 nm, and the visible range is approximately 400 to 800 nm. Upon exposure to light, molecules will absorb or transmit light, depending on the chemical composition of the irradiated material. A specific spectrum is generated with a specific wavelength corresponding to the characteristic functional group of the scanned biocomposite [16]. Factors affecting spectral characteristics are the chemical composition and sample light scattering properties related to its microstructure. In order to use this method, the sample should have functional groups that can absorb light in the UV–VIS region, for example, an aromatic ring. This method also offers the rapid monitoring of changes in the biocomposite when exposed to variable pH and temperatures [17].

Cazón et al. employed the UV technique to optimize the composition of films based on vegetable and bacterial cellulose combined with chitosan and polyvinyl alcohol (**Figure 2**). In addition, with the application of mathematical and statistical,

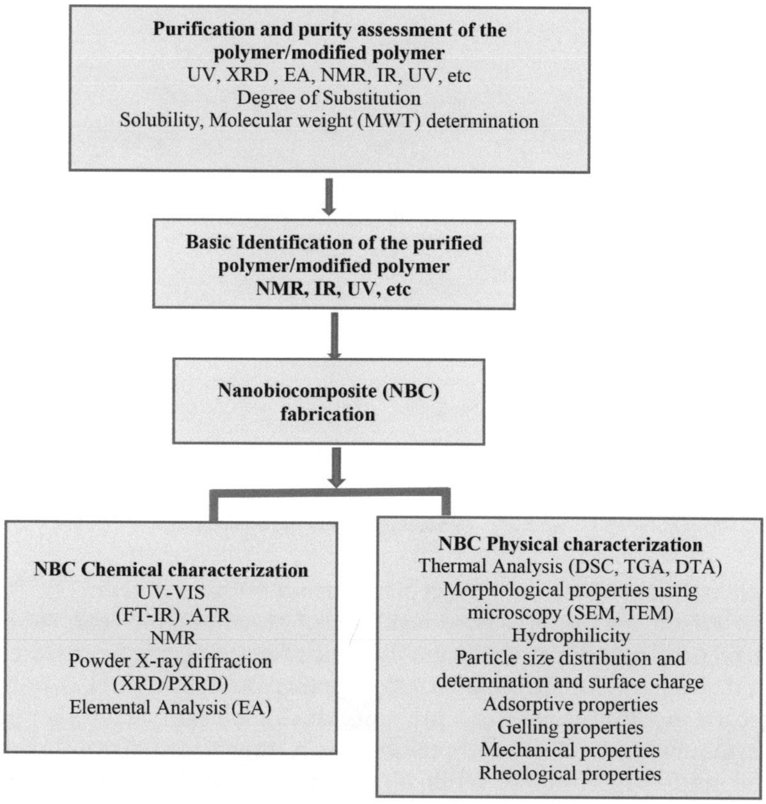

Figure 1.
A proposed approach for the physicochemical characterization of natural polymers and their respective nanobiocomposites (NBC).

they were able to quantitatively estimate the composition of each component with high accuracy based on the data generated from UV–VIS–NIR spectra. Interestingly, the proposed method enabled the discrimination of the geographic origin of the investigated biopolymers [16].

Niroomand et al. used UV–visible spectroscopy to measuring the optical transmittance and opacity of the pure cellulose and Nano-chitosan/cellulose films, by scanning at 200–800-nm wavelengths using Eq. (1) [18]:

$$Opacity = 1/4\ \frac{Absorbance\ at\ 600\ nm}{Film\ thickness\ (mm)} \tag{1}$$

A high transmittance indicates film transparency. Additionally, the researchers observed a slight increase in the film opacity at high dosing rates reaching 15% of nano-chitosan particles, which can result from partial agglomeration of NBCs [18].

2.2 Fourier transform-infra red (FT-IR) and attenuated total reflection (ATR)

In order to use Infrared spectroscopy (IR) as a characterization tool, NBC molecules must absorb light in the infrared region of the electromagnetic spectrum, converting it to molecular vibration. This absorption is measured as a function of

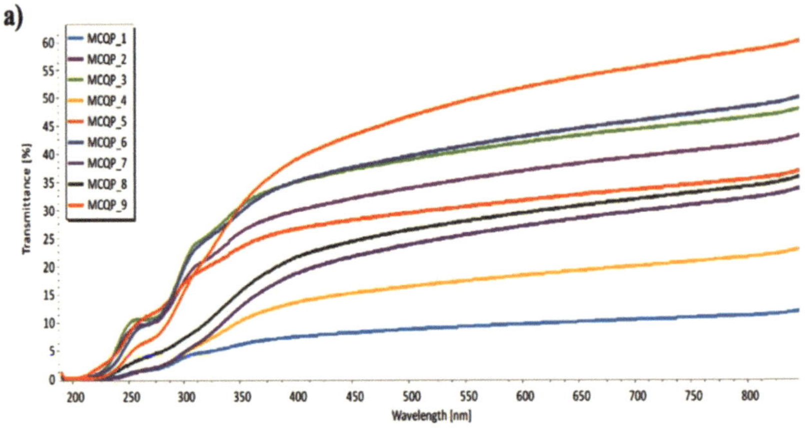

Figure 2.
UV–VIS region spectra of the bacterial cellulose samples with chitosan and polyvinyl alcohol. MCQP is variable composition of bacterial cellulose with chitosan and polyvinyl alcohol [16].

wavelength (as wave numbers, typically from 4000 to 600 cm^{-1}). This absorption is characteristic of a sample's chemical bonds (stretching, binding, etc.) [19]. Using a mathematical algorithm, the wave number raw data is transformed into an IR spectrum that serves as a characteristic "molecular fingerprint" that can be used in the structural identification of organic samples. A solid sample is either ground with IR potassium bromide (KBr) and pressed into a transparent disc or is thinly sliced and placed onto a KBr window. While liquid samples are directly measured or diluted with an IR transparent solvent [19, 20].

Other IR techniques based on reflection rather than transmittance are Diffuse Reflectance Infrared Fourier Transform Spectroscopy (DRIFTS)and Attenuated Total Reflection (ATR). In DRIFTS, the IR light only interacts with the surface of a material to collect chemical information following mixing with KBr. Besides being a non-destructive method, ATR is a simple handling technique that utilizes a limited amount of the tested sample that is directly placed in a zinc selenide (ZnSe) crystal or diamond without any other ingredient. It can be used to study soft, stiff, and rigid polymers. However, ATR has the drawback of generating false data. ATR interrogates the surface of the sample so that the chemistries on the surface are amplified. The refractive index determines the strength of the reflection, so wherever an absorption band is present, the extent of the reflection will change [21]. Generally, it is essential to examine the unmodified polymer and each NBC component separately before scanning the NBC itself. This allows the observation of new bands, changes in band intensity, etc., that can be characteristic of the NBC. In addition, caution should be made to avoid non-uniform particles or large particles that can affect the data generated in the DRIFTS and ATR. Some disadvantages include the interference of water, CO_2 effect, etc. [22].

IR has been widely applied in measuring grafted polymers' functionalization prior to NBC fabrication [23]. **Figure 3** shows the IR spectrum of acrylic functionalization of cellulose nanocrystals with 2-Isocyanatoethyl Methacrylate(IEM). Strong absorptions at 1723 and 1640 cm^{-1} indicated the attachment of C=O and C=C groups on modified cellulose nanocrystals (mCNCs). The disappearance of the –NCO peaks from 2-isocyanatoethyl methacrylate and the appearance of the multiple absorption peaks between 1200 and 1700 cm^{-1} is associated with the formed urethane linkage

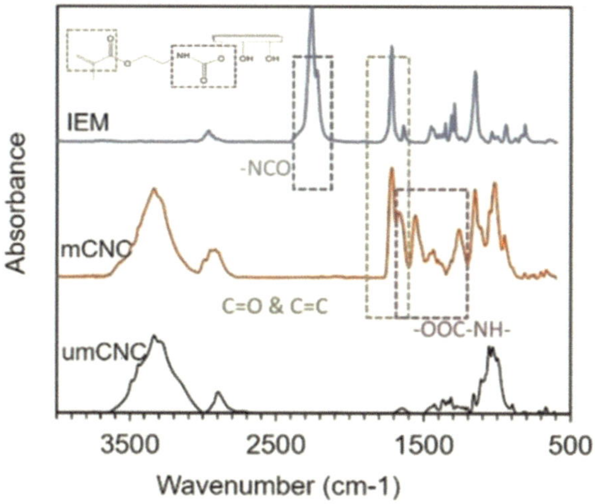

Figure 3.
ATR-FTIR spectra of Unmodified Cellulose nanocrystals (umCNC), modified Cellulose nanocrystals (mCNC), and 2-isocyanatoethyl methacrylate (IEM) [24].

between cellulose nanocrystals and IEM. A small increment of the absorption between 2900 and 3000 cm^{-1}, and the increased C—H stretch of the methyl group of IEM indicates functionalization [24]. Similar studies that reported the use of IR in confirming functionalization include alginate [25], methoxylated pectin and amidated LMP [13], chitosan functionalization with the PNIPAAm [26], cellulose-propargylated chitosan [27], Melamine-Functionalized Chitosan: [28], etc.

IR can be used to study crosslinking within NBCs. For example, Stanescua et al. reported using FT-IR to study crosslinking in NBC based on chitosan/bacterial cellulose used as a wound dressing. FTIR spectra show three new peaks in the regions 2980, 2972, and 1425 cm^{-1}, which can be attributed to the asymmetric stretching vibration of C—H from methyl groups and for bending vibration of C—H from methyl groups of the grafted chitosan, respectively (**Figure 4**). These vibrations suggest the generation of a new physically crosslinked network via hydrogen bonding [29].

In addition, IR can be used to confirm the immobilization of biomolecules onto NBCs by observing the new bands or changes in the IR band pattern representing specific chemical groups. İlgü et al. investigated the immobilization of recombinant esterase onto chitosan nanoparticles (NPs) by physical adsorption under several immobilization conditions. As seen in **Figure 5**, the chitosan nanoparticles (NP) and the enzyme immobilized chitosan NP spectra ((C) and (D)), where the peak intensity at 1650 cm^{-1} has increased, while two peaks shifted from 1560 to 1550 cm^{-1} and from 1413 to 1407 cm^{-1}. The strength of these two peaks' intensity also decreased dramatically. These changes in the FTIR spectrum confirmed the immobilization of esterase on chitosan NPs [30].

2.3 Nuclear magnetic resonance (NMR)

This technique is widely used in identifying and characterizing novel functionalized polymers and their respective NBCs. It gives some idea about the NBC chemical structure, and morphology, for example, the amount and orientation of crystalline phases in semi-crystalline NBCs and the domain sizes in phase-separated polymeric NBCs [31].

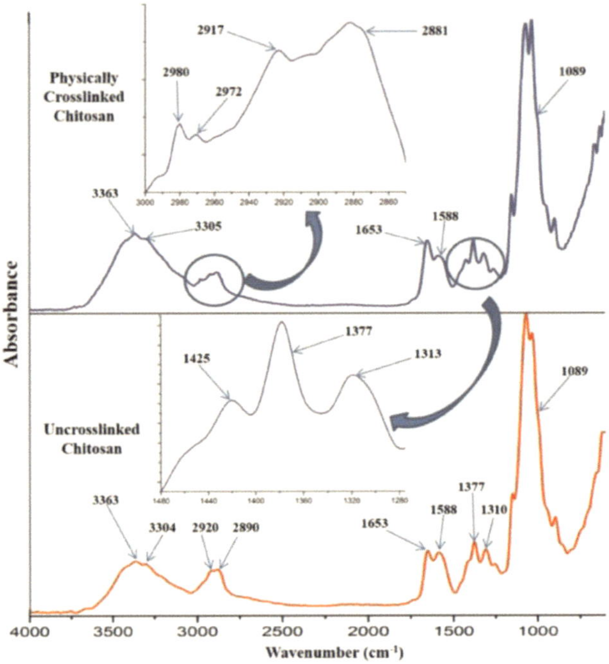

Figure 4.
FTIR spectra of uncrosslinked and physically crosslinked chitosan (individual component) [29].

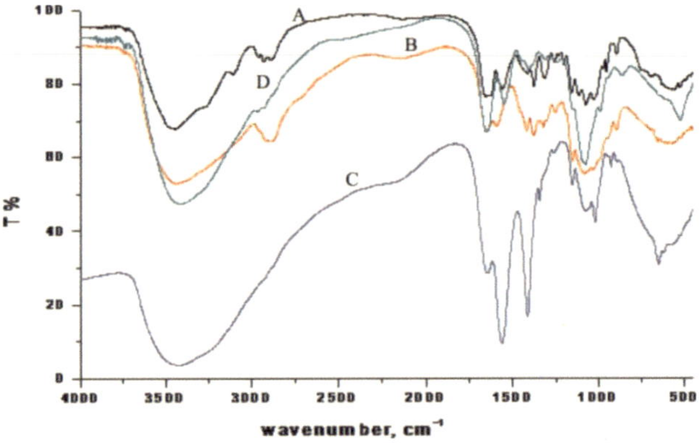

Figure 5.
FT-IR spectra of (A) chitin, (B) chitosan, (C) chitosan NP, (D) enzyme immobilized chitosan nanoparticles [30].

NMR is based on exposing a charged nucleus like hydrogen (^{1}H NMR) or carbon (^{13}C NMR), etc. to a strong magnetic field, which allows the transfer from a low energy state to a high energy state corresponding to a certain radio frequency. The energy is then emitted at the same frequency when the spin returns to its base level. Capturing these signals will give an NMR spectrum with characteristic chemical shifts for the spinning nucleus [32]. The precise resonant frequency of the energy transition is affected by electron shielding, which in turn is dependent on the chemical environment (i.e., the functional group within a polymer). The presence of an electronegative group around the nucleus will result in a higher resonant frequency in general. In order to get accurate measures using this technique, the following factors should be considered: signal-to-noise ratio, saturation effects, peak shape, resolution, isotopic satellite, spinning sidebands, baseline slant, and curvature [33, 34]. The advancement of NMR to include computer-assisted methods enabled more information on molecular modeling and conformational analysis of many natural polymers [33].

NMR can be used to characterize liquid NBCs samples like gels, dispersion, melt, and solutions with increased spectral specificity compared to solid samples [31, 32]. Consequently, dilution, dispersion, increased temperature, etc., can give more information related to polymeric microstructure, its dynamics, and interactions with other ingredients within an NBC [33]. Liquid state NMR can be classified into one-denominational and multi-dimensional (2D and 3D) techniques. Hyphenated techniques include On-line HPLC-NMR, Supercritical Fluid Chromatography (SFC)-NMR, and Offline capillary electrophoresis. Isotopic labeling (^{13}C, ^{19}F, ^{15}N, and ^{31}P) can facilitate relaxation studies of polymers hence, enabling studying crystallization within these polymers. Nano-NMR can be used to study heterogeneous and limited quantity samples [31, 35].

On the other hand, the solid-state NMR spectrum tends to have broad lines because of chemical shift anisotropy and dipolar and quadrupolar couplings. High-power dipolar decoupling, cross-polarization, and magic angle spinning to produce high-resolution ^{13}C NMR spectra avoid long instrument running time. Likewise, the combined rotation and multi-pulse (CRAMPS) experiment can permit H spectra with narrower line widths to be obtained [31, 33, 35].

NMR has been extensively used to determine the degree of functionalization of many natural polymers and to characterize their corresponding NBCs through careful measurement of peak heights or areas under the signal peak in the NMR spectrum using a suitable reference standard. Examples of published literature about using ^{1}H NMR include: include N-carboxyethyl chitosan and glycol chitosan [36], 2,3-epoxypropyltrimethylammonium chloride grafted starch [37], heteroaryl pyrazole chitosan derivatives [38], sulfonated chitosans [5], Cellulose Nanocrystals with 2-Isocyanatoethyl Methacrylate [24] and many others. 1H NMR has also been used to predict the rigidity of polymers and different phases of a polymer like cellulose [33]. The following paragraphs will discuss some of these published data.

Zhou et al., 2022 et al. used NMR spectroscopy to study the functionalization of chitosan and to detect the suitable pH that enables optimum functionalization (**Figure 6**). The location of the chemical shift of the aliphatic portions of the chitosan and the aromatic protons of the grafted substituent is a confirmation of the grafting procedure. Interestingly, signal intensity at δ 6.823 ppm (typical aromatic proton signal of substituent) was dependent on the pH of the medium, i.e., higher substitution was observed at pH 6.4 compared to pH 3.4 [39].

NMR can also be used to monitor the start and end of the gelation process in NBCs and to predict stability over time. Craciun et al. reported the preparations of a high water content chitosan-based hydrogel that was monitored in deuterated water over 22 days at room temperature (**Figure** 7) Gelation was driven by the formation of an imine group between chitosan (NH_2 group) and vitamin B6 precursor, pyridoxal 5-phosphate (aldehyde group). The beginning of the gelation process was evidenced by the appearance of the chemical shifts of the imine group, while the progressive diminishing of the integrals of imine and aldehyde protons and the appearance of the enol proton (around 6.5 ppm) indicated the end of the process. This suggests that a too diluted system favored the shifting of imination to the reagents and the stabilization of

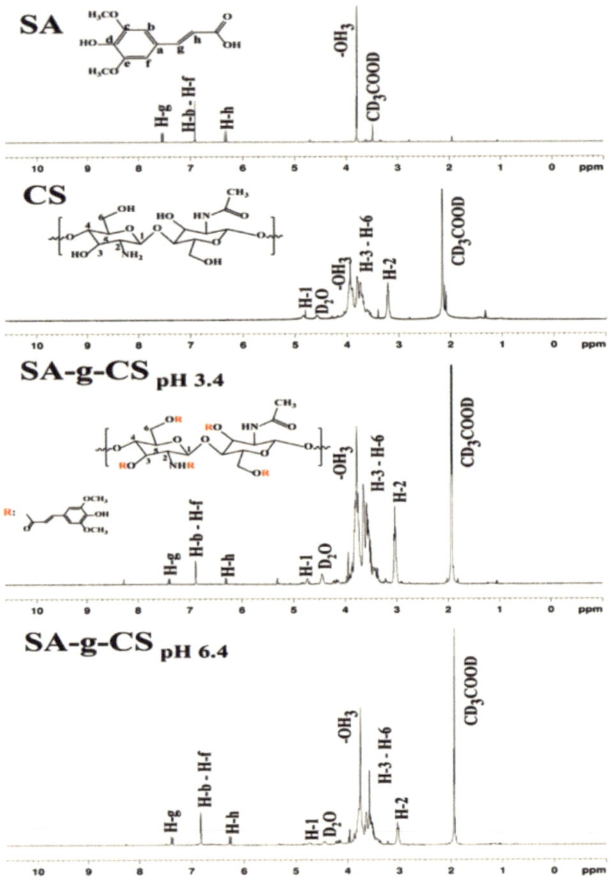

Figure 6.
1H NMR spectra of sinapic acid (SA), chitosan (CS) and sinapic acid-graft-chitosan (SA-g-CS) conjugates synthesize under different pH conditions (3.4–6.4) [39].

the enol form of aldehyde. Consequently, the water content of this hydrogel can limit the storage duration at room temperature to less than 22 days [40].

Heinze et al. used ^{13}C NMR to identify the functional groups following grafting 2,3-epoxypropyltrimethylammonium chloride to starch. The ^{13}C NMR in **Figure 8A** shows different carbons type in the backbone of the starch and its grafted conjugate. Additionally, the researchers used Distortionless Enhancement by Polarization Transfer (DEPT) NMR to confirm the connections between the carbons. DEPT can be easily combined with a 1H isotropic-chemical-shift filter that selects NCH/OCH signals versus CCH(C, C) signals (**Figure 8B**) [37].

On the other hand, 2D NMR like COSY (COrrelated SpectroscopY, H-H NMR) and HSQC (cross-polarization heteronuclear single-quantum coherence, H-C NMR) can be used to study correlations between two nuclei which are separated by one bond like two hydrogens or hydrogen and carbon within a chemical structure. Since all NBCs have carbon and hydrogen atoms, either in the grafted molecules or in the host structure itself, 2D-NMR can be a valuable tool for studying the structural interactions in NBCs [32]. The COSY technique involves plotting ^{1}H NMR for each component to detect the proton-proton interaction, which is then plotted as 2D contours in the XY plane to detect interaction dynamics. Wang et al. used 2D-NMR (COSY) to study

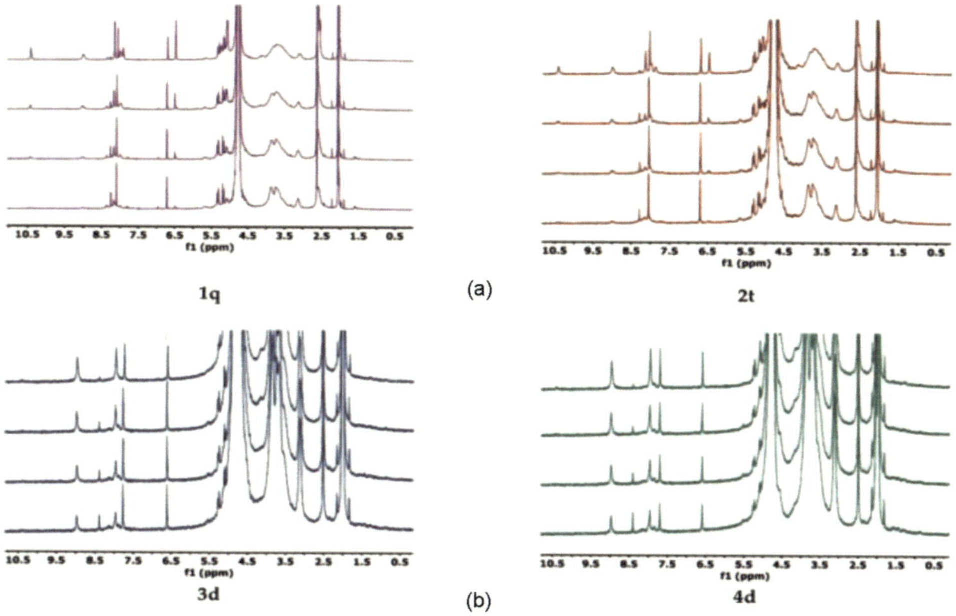

Figure 7.
Representative 1H-NMR spectra of hydrogels (a) with higher water content 1q, 2 t and (b) lower water content 3d, 4d, recorded over time (from up to down: 1, 7, 15, and 22 days) [40].

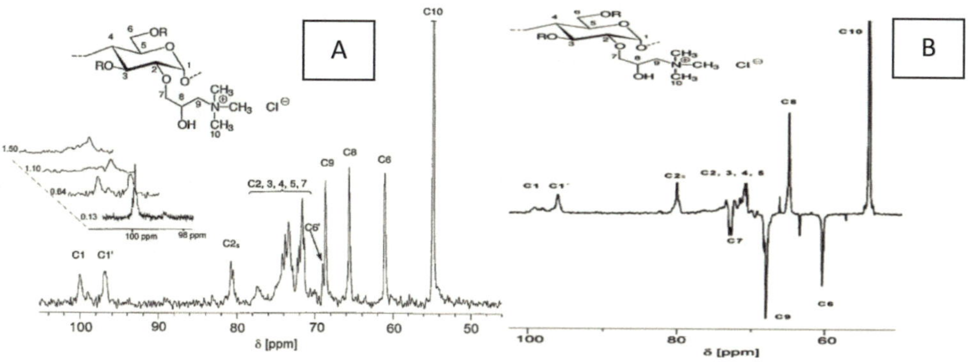

Figure 8.
(A) The 13C-NMR spectrum of cationic starch sample H 6 (degree of substitution (DS) =0.66) was measured in D2O at 607C (R = H or cationic group according to DS). (B) DEPT-135 spectrum of cationic starch P 3 (degree of substitution = 1.05) [37].

drug-polymer interactions of N-succinyl chitosan-alginate grafted NPs loaded with mangiferin (an anti-atherosclerotic drug). As seen in **Figure 9A**, the drug and N-succinyl chitosan interact at several neighboring protons; for example, the phenolic –OH of mangiferin (δ 9.25) interacts with the –COOH proton of the succinyl side chain (δ 12.38).and the –OH of mangiferin and –C—O, –NH–, carboxyl of side chain –NHC– – OCH_2CH_2COOH in N-succinyl chitosan [41]. On the other hand, HSQC (^1H—^{13}C) can also be used to study the interaction between hydrogen and carbon atoms within a biopolymer. Huamani-Palomino et al. utilized the HSQC spectrum to confirm the purification process of alginate (**Figure 9B**) by observing the coupling bands of the alginate monomers (glucoronic acid and mannuronic acid) [25].

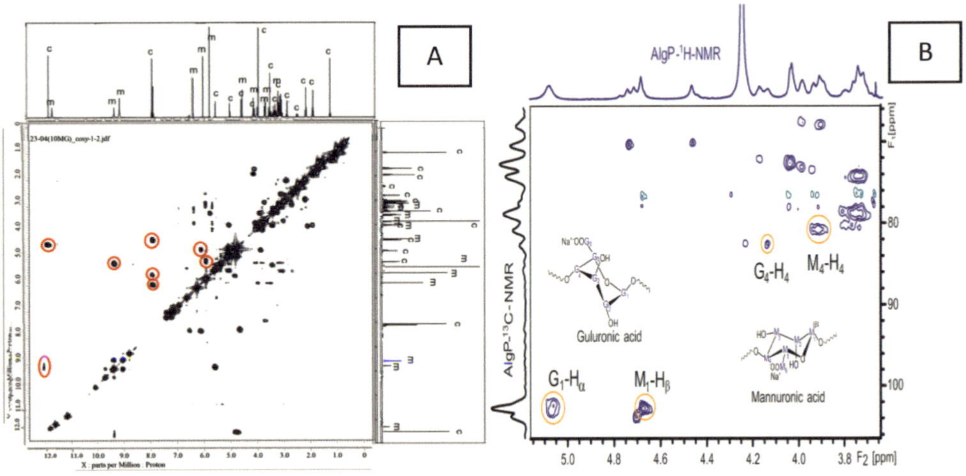

Figure 9.
(A)2D NMR (COSY) spectra of mangiferin loaded NSC-alginate formulation. C-denotes proton peaks for N-succinyl chitosan and m-depicts proton peaks for mangiferin [41]. (B) HSQC spectrum of purified alginate (AlgP) [25].

2.4 Powder X-ray diffraction (XRD/PXRD)

It is well established that natural polymers show a variable proportion of their disordered amorphous regions and ordered crystalline regions, which consequently affect the characteristics and applications of their NBCs. The presence of amorphous regions highly affects the polymer plasticity and flexibility, while the crystalline regions affect the elasticity and stiffness of these materials [42]. The amorphous/ crystalline proportion of a natural polymer and their respective NBC are greatly affected by purification and drying (solvent evaporation, lyophilization, etc.) [43]. X-Ray diffraction (XRD) is an analytical technique that is used to determine the solid state, crystal size and shape, and phase identification with quantitative phase analysis of materials. The theoretical basis of X-Ray diffraction stands on Bragg's Equation (Eq. (2)) [43]:

$$n\lambda = 2d \sin\theta \tag{2}$$

Where n is the order of reflection n = (1, 2, 3, ...) λ, the wavelength, d the distance between parallel lattice planes, and the angle between the incident beam and a lattice plane, known as Bragg angle [44]. The geometry of the crystal lattice determines the position of the peaks in an X-ray diffraction pattern. In general, as the material became more symmetrical, the peaks became fewer in its diffraction pattern. The peak intensities associated with the diffraction intensity are determined by the arrangement of atoms within the crystal lattice [45].

Experimentally, there are two methods of XRD, the Laue method, where θ is kept constant and λ varied, and the powder diffraction method, where λ remains constant, and θ is varied. In both methods, the intensity of the diffracted X-ray beam against diffraction angle 2θ is measured, which gives the diffraction pattern of the material. The pattern obtained in crystalline materials shows sharp maxima, called peaks, at their respective diffraction angle, and in amorphous solids, the orderly structure is absent, which gives rise to broad maxima called a hump [42]. X-ray scattering provides structural information at three different length scales by performing scattering experiments at such as 1 (wide XRD), 10 (small XRD), and 100 nm (ultrasmall XRD) angles. Natural polymers in general are not fully crystalline; so XRD is used to measure their degree of crystallinity [45]. Three important information are needed when interpreting an XRD diffractogram:

1. The position of the diffraction peaks.

2. The peak intensities.

3. The intensity distribution as a function of diffraction angle.

Prior to performing any XRD measure for an NBC, it is essential to scan each component alone (drug alone, polymer alone, crosslinker alone, etc.) followed by a scan of the physical mixture of two or more components, and finally, the NBC in order to compare the molecular interaction (**Figure 10**) [6]. Any peak position or intensity change indicates an interaction between the drug and the polymer upon NBC fabrication. At the same time, the broadening of peaks (halo-pattern) or decreased intensity of a peak indicates amorphous transition or the presence of an amorphous state [45].

Drug-polymer interaction within NBCs can also be studied by detecting the presence or absence of new peaks. For example, the XRD of mebeverine(MB) loaded chitosan NPs shows broad peaks indicating an amorphous state within the polymer or

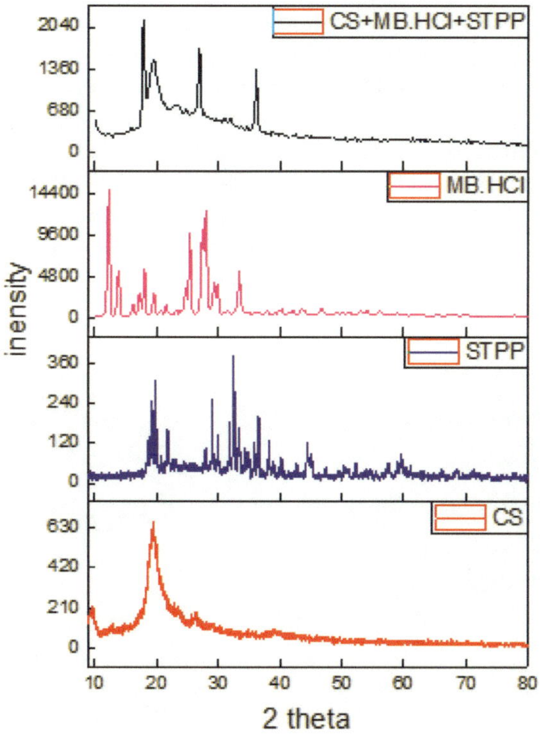

Figure 10.
XRD spectrum for chitosan polymer (CS), crosslinker (STPP), Mebeverine Hydrochloride(MB.HCl), and Chitosan nanoparticles loaded with MB.HCl (CS + MB. HCl + STPP) [6].

the NP (**Figure 10**). Moreover, the absence of additional peaks indicates the purity of the formulations. Bragg Law was used to calculate the crystallization of the chitosan polymer crystal practices, and it reached 4.5. Upon comparing the XRD of the NPs and that of each component, it can be concluded that the peak at 2Θ = 26.9o is due to the sodium tripolyphosphate (STPP) crosslinker-drug interaction, while the peak at 2Θ = 17.9o is related to the loaded drug mebeverine. Additionally, the absence of any additional peak indicates no change in the degree of crystallinity of the polymer during the fabrication of the NPs [6].

Similarly, Shahid et al. used XRD to predict the type of interaction within Ticagrelor-loaded chitosan-based NPs [46]. Ahmad et al. reported the use of XRD to study the crystallite structure within starch-based NPs prepared using mild alkali hydrolysis and an ultra-sonication process. Using the quantitative measurement of the area under the amorphous region and diffraction peaks the researchers concluded a decrease in crystallinity. The increased amorphous region was accompanied by diminished diffraction peaks following the size reduction of starch to the nanoscale [44].

The stability of NBCs during storage can also be assessed using XRD through the evaluation of their crystallinity over time. Burapapadh et al. evaluated the degree of crystallinity of itraconazole (ITZ) pectin-loaded NPs. **Figure 11A** shows that pectin alone exhibited a halo pattern indicating the amorphous state of the polymer, while the sharp peaks (17.45 and 17.95 (doublet), 20.30, and 23.45 2θ) of the drug alone suggest its high crystalline nature. The XRD patterns of drug-polymer physical mixtures showed similar peaks as untreated drug, indicating no change in drug crystallinity during the mixing process, while the XRD patterns of NPs showed the absence of the characteristic crystalline drug peaks (a typical broad hump of amorphous material), indicating that the drug is present on the noncrystalline form within the NPs. Likewise, the assessment of the XRD of the prepared sample after one-year storage at 25°C, showed (**Figure 11B**) the halo-pattern of the molecularly dispersed amorphous drug. However, there were some crystallinity peaks presented at approximately 12 and 21 2θ degrees. This indicates the start of transformation from amorphous to crystalline solid upon storage of the NPs for 12 months [43].

Small-angle X-ray scattering (SAXS) method utilizes smaller angles in scanning, typically from 0.1 to 10°, where the elastic scattering of X-rays caused by nanoscale

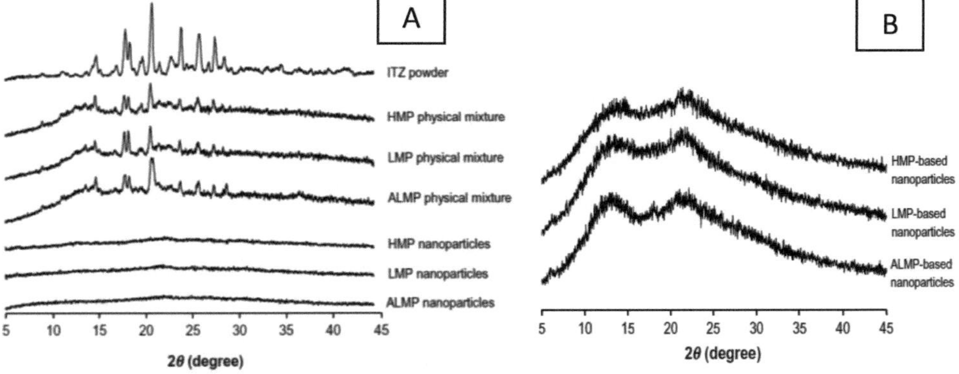

Figure 11.
(A) Powder X-ray diffraction patterns of ITZ, physical mixture of ITZ, and various types of pectin and nanoparticles prepared from nanoemulsion templates. (B) Powder X-ray diffraction patterns of various nanoparticles prepared from nanoemulsion templates, using a mechanical homogenizer, after 1-year storage at ambient condition (25°C) [43].

structures in the polymer is recorded. This enables studying NBCs in the range of 0.5–100 nm up to 1000 nm. Such information includes size, shape, pore sizes, and other characteristic distances of partially ordered materials collected [36, 42]. Lin Y et al. reported sing SAXS coherent X-ray scattering (CXS) in studying the dynamics and gelation mechanism of glycol and carboxyethyl chitosan-based hydrogels with different dynamic interactions. In situ SAXS enables getting information about the nucleation and growth mechanism during the gelation process to form a hydrogel. Moreover, the continuous time-resolved CXS profile unveiled the dynamic behavior of different self-healing hydrogels in mesoscale, supported by rheological experiments [36].

2.5 Elemental analysis (EA)

This method is based on determining the molecular compositions by calculating the ratios of each element within a polymer, for example, carbon, nitrogen, hydrogen, etc., in addition to halogens. In the elemental analyzer, the sample is combusted at 1000°C in a special furnace. The analysis is accomplished by the quantities of CO_2, H_2O, and NO_2 produced by the combustion of the dried carbonaceous materials in excess oxygen. The weights of these combustion products are used to calculate the combustion of samples. The weight percentage of C and H is determined by infrared detection, whereas N content is measured by thermal conductivity detection [24, 47].

Li et al. reported using EA to detect the functionalization of glucose-conjugated chitosan nanoparticles (GCNPs). EA was used to determine the percentages of C, H, and N and the degree of N-succinyl glucosamine substitution (DS). The method is based on calculating the percentage ratio of the atomic mass of C, H, and N to the substituted and the unsubstituted chitosan [48].

One of the advantages of EA is the small amount of sample to be tested (5 mg). However, the sample should be as pure as possible and completely dry. Any impurities or trace solvents will interfere with the results and make the interpretation difficult. Ideally, solid samples should be tested in powder form. The analysis should be carried out under the nitrogen gas purge for air-sensitive samples.

Sample	Reagent	Content (% m/m)			Substitution (mol%)
		N	*C*	*H*	Degree of amidation
1		0	32.85	5.20	—
2	methanol	0	38.01	5.39	—
3	*n*-butylamine	4.58	46.00	6.95	77.86
4	*n*-hexylamine	4.07	49.83	7.59	72.34
5	*n*-octylamine	4.08	52.98	8.20	83.96
6	*n*-dodecylamine	2.87	54.94	8.77	57.94
7	*n*-octadecylamine	2.62	58.11	9.12	76.55
8	ethylendiamine	7.79	34.02	6.24	73.24
9	ethanolamine	5.37	38.84	6.29	93.19
10	hydrazine	6.21	29.84	4.90	53.52
11	hydroxylamine	2.52	34.94	5.35	37.15

Table 1.
Organic elementary analysis of the alginic acid and its derivatives [55].

Elemental analysis has been widely used to estimate the degree of functionalized pectin with Polyacrylamide [49], alkyl pectin [50], histidine-pectin [51]. Elemental analysis was also used to characterize cellulose grafted polymers: phenylacetic acid and hydrocinnamic acid [52], 2-propynoic acid, 4-pentenoic acid, 2-bromopropionic acid, or 3-mercaptopropionic acid [53], acryloyl cellulose [24]. Chitosan grafting with poly N isopropyl acrylamide [26], Melamine [28], 5-nitroisatin [54], mono- and di- sulfonic [5], maltol and ethyl maltol [38], cellulose beads [27], Alginic acid with cysteine [25], N-alkylamides, hydrazide, and hydroxamic acid [55]. Starch with 2,3-epoxypropyltri-methylammonium [37], 3- chloro-2-hydroxypropyl) trimethylammonium chloride [56], poly(methyl methacrylate-co-styrene [57]. Taubner et al. reported the elemental analysis of amidated alginic acid. **Table 1** shows the content % of carbon, nitrogen, and hydrogen in addition to the degree of amidation of the various samples [55].

3. Physical characterization

3.1 Thermal analysis

These are a group of methods that examine changes in a solid sample when heated as a function of temperature and time. Information that can be obtained includes crystallinity (melting point), amorphous state (Glass Transition (Tg)), the heat of reaction (enthalpy (H)), thermal stability/degradation, etc. In this section, the most common thermal methods used in NBCs characterization will be discussed: Differential Scanning Calorimetry (DSC), Thermogravimetric (TGA), and Thermal Mechanical Analysis (TMA). It is well established that many features within a thermogram indicate certain transformations within NBCs as described in the following paragraphs [58].

3.1.1 Differential scanning calorimetry (DSC)

DSC determines the solid transitions as a function of temperature and time. It is used to identify solid–solid transitions (crystallization, polymorphism, etc.), melting, decomposition, and others [58]. In general, an initial transition that is observed in DSC is the solvent evaporation while the final thermal peaks can be due to polymer decomposition [58, 59]. Solid-state transitions can be detected following cooling of the sample by re-running the thermal analysis provided the sample is stable with no signs of degradation (change in color, gas evolving, etc.). Crystallization is a kinetic process that is detected by an exothermic peak in a DSC thermogram. The endothermic (heat absorption) and exothermic (heat released) peaks and magnitudes indicate the thermal phase transformation of the composites. The principal thermal data extracted from this analysis are the glass-transition temperature (Tg), degree of crystallization (Xc), crystallization temperature (Tc), and fusion temperature (Tm). The enthalpy variation and heat capacity of the composite can also be determined [60, 61].

3.1.2 Crystallinity and amorphous state of NBCs

Many natural polymers exhibit various degrees of crystallinity. DSC can differentiate these degrees by measuring the Glass Transition (Tg) which is the softening temperature characteristic of an amorphous state. This is attributed to the molecular mobility within the solid sample. This transition highly affects solubility, drug release, drug-polymer interaction, stability during storage, and many other physical properties.

Figure 12 shows DSC thermograms of physical mixtures of high methoxyl (HM) pectin or low methoxyl (LM) pectin and amidated LM pectin (ALMP) used to deliver itraconazole(a poorly water-soluble drug) and their respective NBCs. The thermal properties of physical mixtures of itraconazole and various types of pectin at a ratio of 1:6 were compared to those of their respective NPs. The melting peak of itraconazole crystals can be observed in the first three physical mixtures with an endothermic peak at 166–168°C. However, following encapsulation within the NP, this peak disappeared, indicating molecular dispersion of the drug within the polymer [43].

Xiao et al. utilized DSC to evaluate a series of novel cellulose esters containing phosphorus, including cellulose diphenyl phosphate (C-Dp) and cellulose acetate (CA)–diphenyl phosphate mixed esters. **Figure 13(A)** depicts the Tg of the various grafted polymers compared to native cellulose, which does not exhibit a glass transition when subjected to heat. The thermogram shows a decrease in the Tg value with increasing the degree of substitution of cellulose which was attributed to disrupting the hydrogen bonding in the cellulose hydroxyl groups affecting molecular mobility within the cellulose chain [62].

Likewise, Zheng et al. evaluated the Alkyl pectin with various fatty acid (C4–C16) bromides using DSC. As shown in **Figure 13B**, with longer alkyl chain lengths

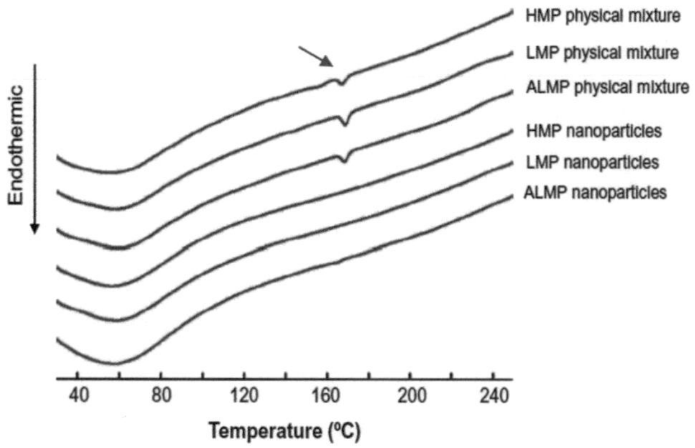

Figure 12.
DSC thermograms of physical mixture of itraconazole and high methoxyl (HM) pectin or low methoxyl (LM) pectin of pectin and nanoparticles [43].

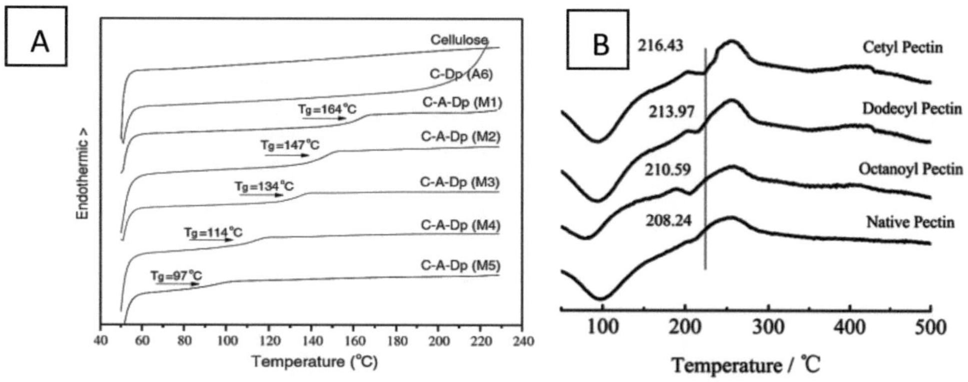

Figure 13.
(A) DSC thermograms of unmodified cellulose, C-Dp (DS = 0.99) and C-A-Dp M1 - M5 with increasing degree of substitution [62]. (B) DSC of unmodified and acylated pectins with different acyl lengths (DS: 10–20%) [50].

(C8—C16), the glass temperature peaks became higher while the peak areas became broader when compared to native pectin [50].

3.1.3 Thermogravimetric analysis (TGA)

TGA is used to study the change in the samples' weight as a function of temperature. Weight loss includes water and solvent evaporation, decomposition, etc. TGA has been widely used to study the effect of varying the type and percentage of nanofillers within an NBC. Polymer decomposition, either in the presence of oxidative or non-oxidative gas, significantly depends on the presence of fillers and their dispersion scale [58]. On the other hand, Derivative thermogravimetry (DTG) is another useful technique that can be used to evaluate the thermal stability of NBCs. Hu et al. reported the use of TGA and DTG to study double-layer hydrogel based on sodium alginate (SA) -carboxymethyl cellulose (CMC) as a sustained drug delivery system. As seen in **Figure 14A** and **B** there are three major weight losses in hydrogels, each corresponding to a change in the nature of the sample. The first weight loss below 100°C is due to water desorption, the second one at 270°C is due to the destruction of glycosidic bonds within the hydrogel, and the third one at about 400°C is due to the destruction of outlayer polymer. Additionally, the incorporation of outlayer polymer (poly(N,N-dimethylacrylamide) (PDMA) or poly (acrylamide) (PAA)) into the hydrogel composition added to the thermal stability and some changes in the degradation pattern. Interestingly, the pronounced delay in the degradation phase (third phase) was attributed to the inclusion of the synthetic polymer (poly(N,N-dimethylacrylamide) (PDMA) or poly (acrylamide) (PAA) in the outer layer of the hydrogel formula [63].

Chang et al. studied the stability of starch-based NPs upon using anionic, cationic, and amphoteric starch NPs. As can be seen in **Figure 15A** and **B**, starch-modified NPs exhibited a lower maximum degradation temperature compared to the maximum degradation temperature of unmodified starch NPs (310.83°C). The results indicated that the thermal stability of modified starch NPs (cationic, anionic, and amphoteric) decreased, evidenced by a lower decomposition temperature, compared to the non-modified starch NPs. This degradation was attributed to the intermolecular forces acting on the starch NPs [56].

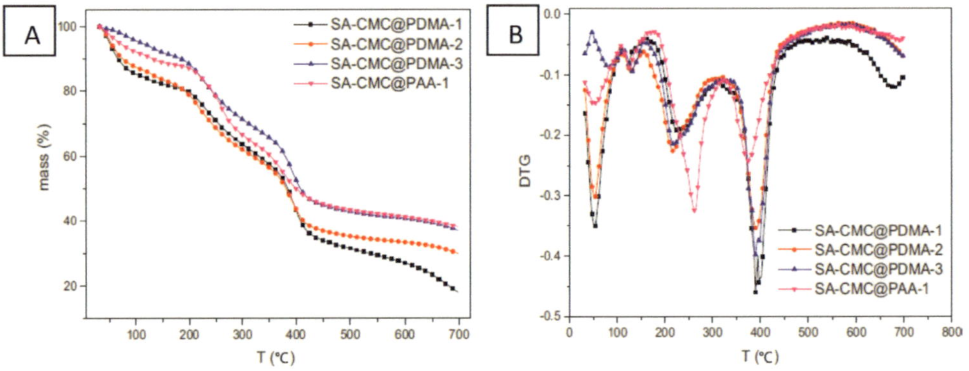

Figure 14.
(A) The TG curves of on NBCs Based on sodium alginate (SA) -carboxymethyl cellulose (CMC) and (poly(N,N-dimethylacrylamide) (PDMA) or poly (acrylamide) (PAA), SA-CMC@PDMA-1, 2,3 and SA-CMC@PAA-1; (B) The DTG curves of SA-CMC@PDMA-1, 2,3 and SA-CMC@PAA-1 [63].

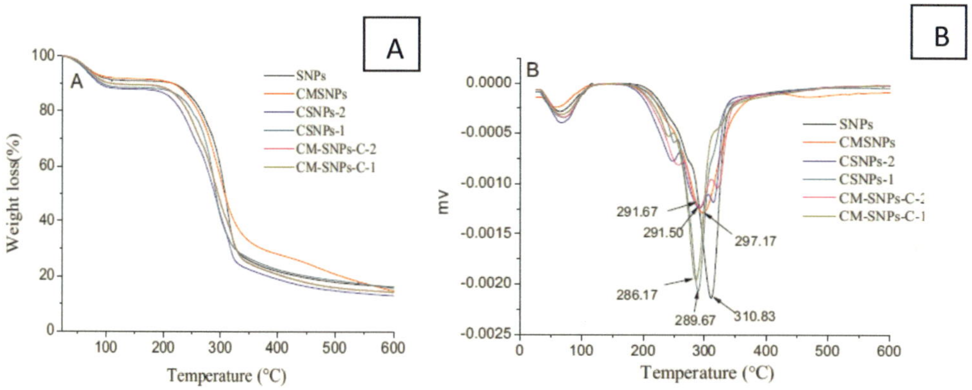

Figure 15.
Thermogravimetry (A) and derivative thermogravimetry (B) of Unmodified starch NPs (SNPs), anionic (CMSNPs), cationic (CSNPs-1, CSNPs-2), Amphoteric (CM-SNPs-C-2, and CM-SNPs-C-1) [56].

Similar findings were reported by Kahdestani et al. during investigating the teicoplanin-loaded NPs based on chitosan/TPP. The TGA was used to evaluate drug and NP degradation over temperature changes. They reported a weight loss of 21.64% occurred between 127 and 226°C related to the removal of the residual water of the NPs and degradation of the polymer (**Figure 16**). Additionally, the authors attributed the increased stability of the NPs to the existence of phosphate groups (P=O and P—O) in chitosan NP leading to less degradation. However, TGA data showed a reduction in the thermal stability of chitosan NPs due to decreased crystallinity within the NPs compared to chitosan alone [64].

In another study, Kassab et al. reported the identification of the various decomposition phases in the range of 120–440°C within NBCs based on sulfuric-acid hydrolyzed cellulose nanocrystal (CNC) extracted from sugarcane bagasse (**Figure 17**). The glycosyl units of cellulose can undergo degradation due to decarboxylation,

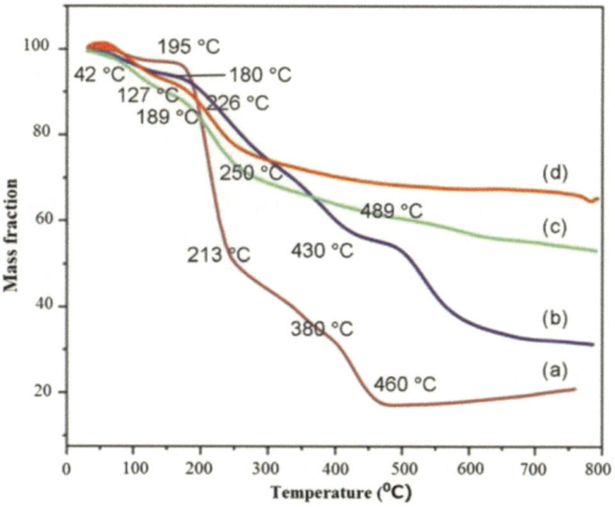

Figure 16.
TGA thermograms for (a) chitosan, (b) chitosan nanoparticles, (c) chitosan nanoparticles containing drug and (d) teicoplanin [64].

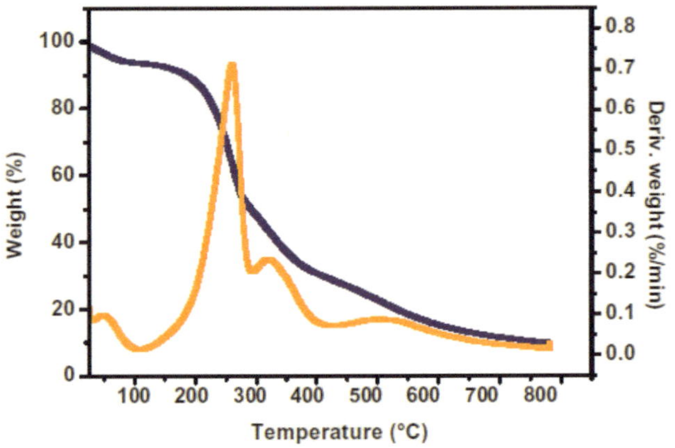

Figure 17.
TGA/DTG curves of sulfuric-acid hydrolyzed cellulose nanocrystal (CNC) [65].

depolymerization, and decomposition. It is reported that the highly sulfated amorphous domains are more sensitive to low-temperature degradation compared to non-sulfated crystalline domains, which are more sensitive to higher temperature decomposition. This impacts the activation energies of the degradation process. The negatively sulfated groups contributed to the decreased thermal stability introduced on the outer surface of cellulose nanocrystal during the sulfuric acid hydrolysis [65].

3.1.4 Thermal mechanical analysis (TMA)

Mechanical properties are very important for thin film evaluation, mainly in the ocular delivery of drugs, since it highly affects drug release rate, swelling, mechanical stability, and other properties [66, 67]. TMA measures the expansion and contraction of NBCs and the effect of crosslinking of the polymers or the enforcing materials [58]. TMA can be used to measure the coefficient of thermal expansion (CTE) of nanocomposite materials which indicates stiffness and energy losses as a function of temperature depending on the degree and the scale of dispersion of nanofillers within NBCs. It also allows the measurement of two different moduli of the nanocomposites, the storage modulus (E′), which is related to the ability of the material to return or store mechanical energy, and the loss modulus (E″), which is related to the ability of the material to dissipate energy as a function of temperature. DMTA data generally showed significant improvements in the storage modulus over a wide temperature range for a large number of polymer nanocomposites [58]. **Figure 18** shows the use of DTMA to evaluate the storage of pectin/cellulose nanocrystal nanocomposite films with varying compositions of (NCC) [68].

3.1.5 Integrated thermal analysis techniques

The integration of thermal analyzers with the microscopy allows visual monitoring of the solid transitions thus, capturing solid-state changes as a function of temperatures and time. For example, Hot Stage Microscopy (HSM) is a combined microscopical technique with DSC/TGA [59]. It has the advantage of evaluating sample morphology, solid state (amorphous, crystalline, polymorphism) transitions, desolvation, and miscibility. **Figure 19** shows the optical micrographs of cellulose

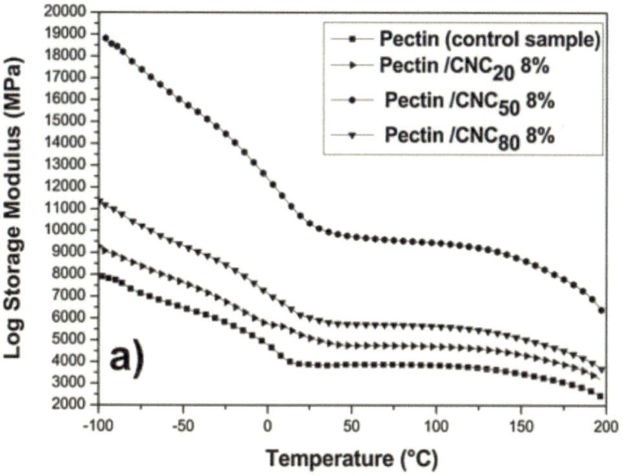

Figure 18.
DMTA curves with Storage modulus (E′) for the neat pectin film and pectin/ CNC20, pectin/CNC50, and pectin/ CNC80 nanocomposite films with 8% filler. (a) [68].

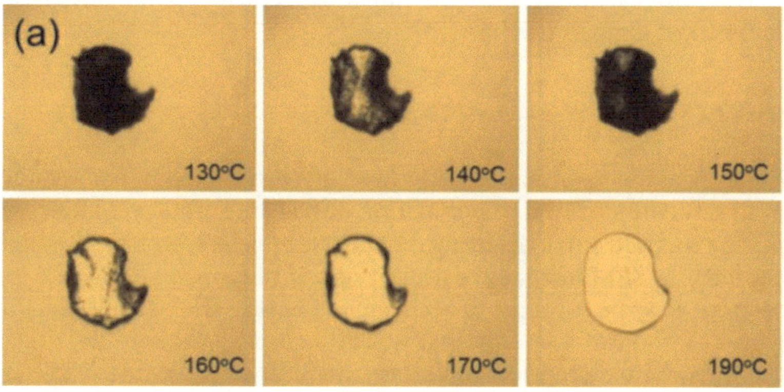

Figure 19.
Optical micrographs of C-A–Dp (sample M5 DSP = 1.19, DSA = 1.53) at different temperatures [62].

diphenyl phosphate at different temperatures. It can be seen that the product started to soften up at 150°C with a complete softening at 170°C [62]. Additionally, HSM can be combined with FTIR and or scanning electron microscopy enabling detailed information about the solid state of NBCs [59].

Additionally, gas evolving during the heating process can be trapped and analyzed using Gas chromatography–mass spectroscopy (GC–MS) techniques. These gases can also be chemically characterized using FT-IR and mass spectroscopy. Additionally, it is possible to combine AFM with DSC, TGA, or any other thermal analyzer enabling the evaluation NBCs [61].

3.2 Morphological properties using microscopy

The size and surface characteristics of NBCs can be acquired by microscopic methods. Optical detection and spectroscopy of a single nano-object can be achieved via detection of NBC interaction with a light beam, i.e., its elastic or inelastic scattering or absorption, or nonlinear ones (such as hyper-Rayleigh scattering or four-wave mixing) [69]. Surface roughness and/ or porousness, homogeneity, diameter, etc., are highly affected by solvent separation, for example, by evaporation resulting in shrinkage and wrinkle formation. Polarized Light Microscopy (POM) is primarily used in NBCs hydrogel evaluation due to their modest sample preparation steps. Craciun et al. investigated the chitosan self-healing hydrogels, designed as carriers for local drug delivery by parenteral administration. Based on the formation of an imine between chitosan and pyridoxal 5-phosphate, the active form of vitamin B6. POM images showed an intense birefringence in a sample of chitosan/pyridoxal 5-phosphate hydrogel, indicating the signature of an ordering degree (**Figure 20**). By coupling X-ray and POM data, it can detect the intermolecular forces that directed a supramolecular arrangement of the imino-chitosan chains [40].

High-resolution methods are now used to get precise dimensions of the NBCs; in addition, they can be used to assess changes over time with regard to agglomeration, swellability, and shear-induced configuration. etc. [29, 70]. NBCs morphology using scattering techniques includes polarized and depolarized light scattering (DLS and DDLS, respectively). Electron microscopy is a technique with a nanometer scale resolution and is capable of imaging NBCs including Transmission Electron Microscope(TEM), Scanning Electron Microscope(SEM), and Atomic Force Microscopy (AFM). Electron microscopy enables the direct observation of the dimensions (i.e., length and width) of a given particle [71].

3.2.1 Transmission electron microscope (TEM)

TEM images provide good nanometric (and often subnanometric) resolution, allowing rapid screening of a large population of particles, thus avoiding major sampling issues. The method involves using a high-energy electron beam to bombard the sample. However, as TEM images are projections of the objects along the incident beam direction, it may be difficult to accurately measure the particle thickness. Depending on the amount of energy that was absorbed by the sample, the intensity of the beam that hits the viewing screen varies, and an image is made [71, 72].

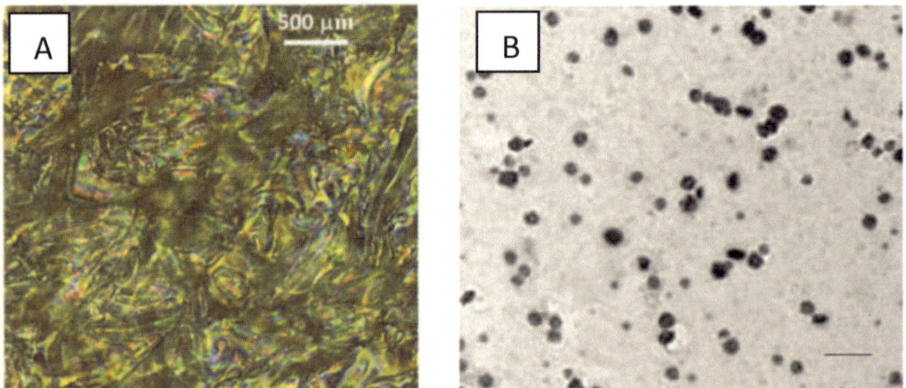

Figure 20.
(A) Representative POM image of xerogel 1.5 t (to be representative, the hydrogels given in the figure have a different molar ratio of glucosamine/aldehyde units and/or different water volumes [40]. (B) Transmission electron microscope (TEM) images of the modified xanthan gum. Nanoparticles [14].

In order to get an accurate measurement using TEM, the sample should be extremely thin (thickness below well below 1 μm) and well dispersed (no agglomerate) to allow transmission of electrons; additionally, the atomic number, density of the observed material, and on the energy of the incident electrons should be considered. Usually, this can be achieved using sectioning techniques for solid samples or preparation of NBCs using dilute suspensions (dispersion) [71]. Then, the sample is deposited on thin circular metallic grids (copper, carbon, etc.) with typical meshes around a few 10 of micrometers and accurately mounted in the sample holder for microscopy. Caution should be taken to avoid degradation of the copper grids when a high or low pH dispersion is investigated. Moreover, to avoid aqueous sample accumulation due to the hydrophobicity of the grids, glow discharge should be performed by placing the carbon-coated grids inside a partly evacuated chamber connected to a power supply. This allows the electron potential to ionize the gas within the chamber where negatively charged ions deposit on the carbon, giving the carbon film an overall hydrophilic (water-attracting) surface [71, 72]. **Figure 20(B)** depicts TEM image of modified xanthan-gum based [14].

NBCs main components, carbon, oxygen, nitrogen, and hydrogen, are not very dense, so sample staining is needed since the number of electrons they absorb is minimal compared to the intensity of the electron beam. Therefore, a heavy metal salt that readily absorbs electrons like lead, tungsten, molybdenum, vanadium, or depleted uranium is usually used. After staining, the sample is blotted, air dried, and ready to be examined in the microscope [71].

3.2.2 Scanning electron microscope (SEM)

SEM analysis technique uses electron selective detection methods capable of nanoresolution and chemical characterization of NBCs [70]. SEM micrographs can be used to study NBC porosity, the uniformity of pores, pore interconnectivity, and their size. NBC porosity can be a measure of the ability of NBCs to swell and deswell, which highly affects drug release mechanisms. **Figure 21** depicts the uniform porousity of hydrogel NBCs based on chitosan (CS), xanthan gum (XG), monomer 2-acrylamido-2-methylpropane sulfonic acid (AMPS) that was successfully used to deliver acyclovir [73]. Pore dimensions are highly affected by the water content and the degree of crosslinking

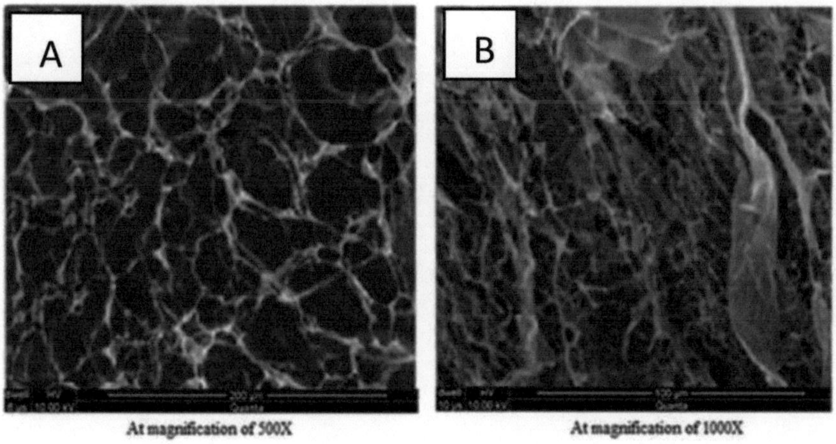

Figure 21.
SEM images of acyclovir loaded hydrogel (A) At 200 μm (B) At 100 μm [73].

within the NBC. Crosslinking highly affects intermolecular physical connections among the chitosan. SEM can provide a detailed histogram describing the change in the number of pores with water content to enable optimization of the dispersion volume of NBCs, as depicted in **Figure 22** [40].

3.2.3 Integrated SEM techniques

SEM- Energy Dispersive X-ray Analysis (EDAX) can be used to study the morphological appearance of the hydrogels. Stanescu et al. Used SEM-EDAX integrated with elemental analysis to detect the proportions of elements like carbon and oxygen in chitosan and bacterial cellulose NBCs for wound dressing (**Figure 23**) [29].

3.2.4 Field emission gun scanning electron microscopy (FEG-SEM)

FE-SEM can be used to study the distribution and cross sections of the nanocomposites within a matrix [74, 75]. Niroomand et al. used FESEM to study the morphology of the cellulosic matrix and NBCs films [18]. **Figure 24** depicts FE-SEM of differences in sample preparation of dried BC nanofiber, or NBCs film prepared by blender (**Figure 24A**) or homogenizer (**Figure 24B**) [18]. The film was coated with gold to minimize the electron charge (**Figure 24C**) [18].

3.2.5 Atomic force microscopy (AFM)

AFM provides information on morphology, surface topography (roughness and transparency), mechanical properties, and adhesion of NBCs. It also provides accurate measurement of NBCs size and size distribution. The high spatial and force resolution of

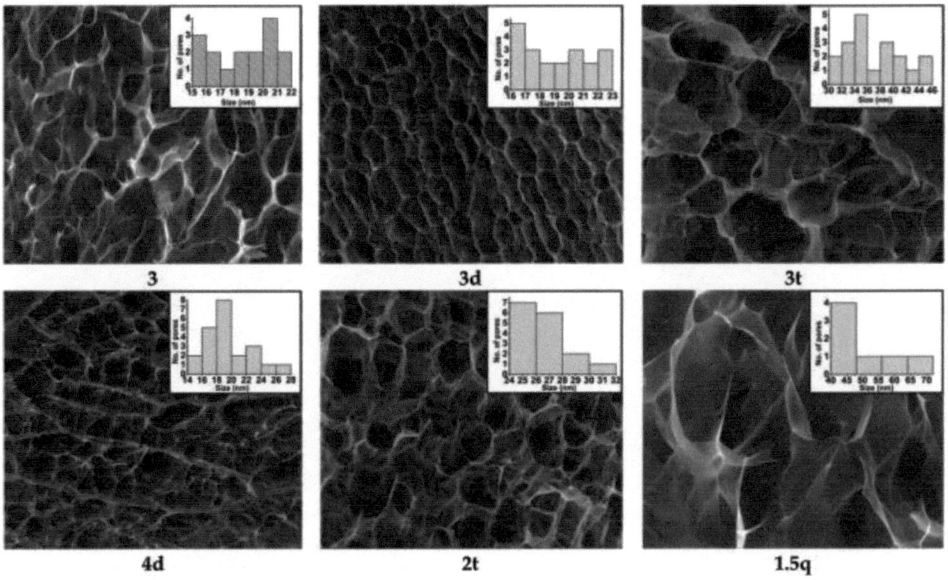

Figure 22.
SEM images for representative xerogels (scale bar: 50 μm) and corresponding histograms (to be representative, the samples given in the figure have a different molar ratio of glucosamine/aldehyde units and/or different water volumes) [40].

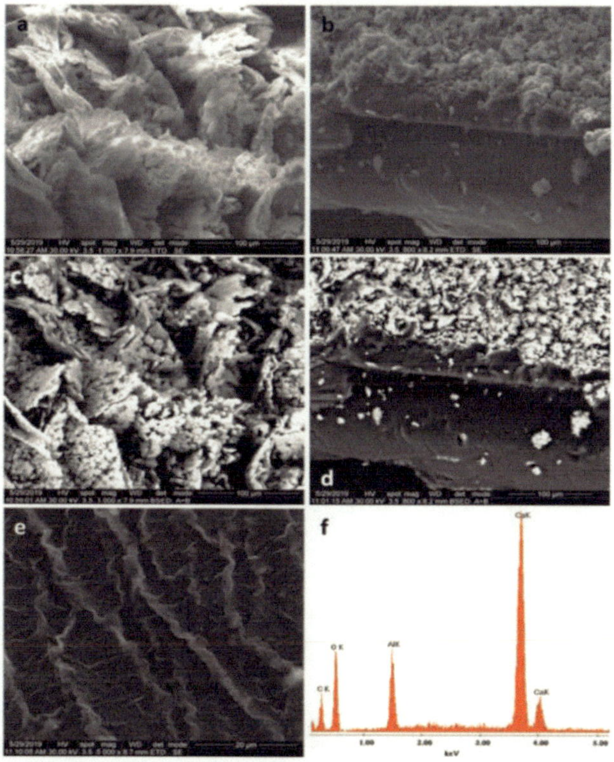

Figure 23.
SEM images of: (a,c)BC/chitosan membrane surface; (b,d); lateral view of BC/chitosan; (e); scattering distribution crude chitosan surface, (f) EDAX elemental analysis [29].

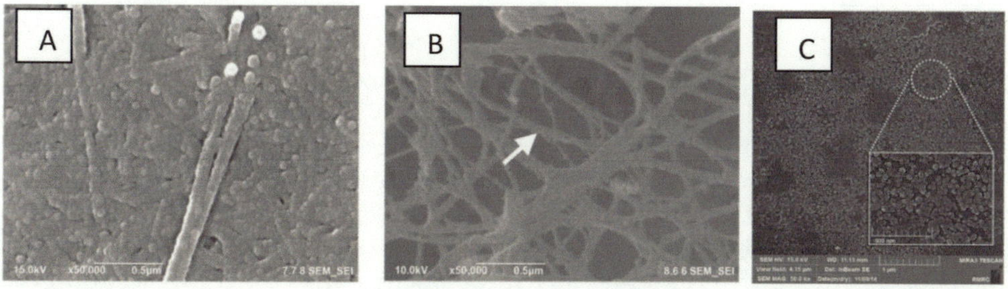

Figure 24.
FESEM morphology of bacterial cellulose nanofibers after treatment with (A) blender and (B) homogenizer [76]. (C) FE-SEM image of the synthesized Nano-CS particles [18].

AFM up to sub-nanometer scale and pico newton forces offers several unique advantages. It can also provide high-resolution mechanical testing along with images [70, 72]. Additionally, it also works with some special in-situ settings, such as imaging samples with fluid layers or wet samples. However, due to the scanning mechanism of AFM, an artifact may be introduced by contamination layers, temperature change, surface shear, or complex surface topography. The AFM scanning in high resolution also requires very small scanning steps. Along with the distance limitation of non-contact or tapping mode, the scanning of relatively large non-smooth areas or complex overall shapes can

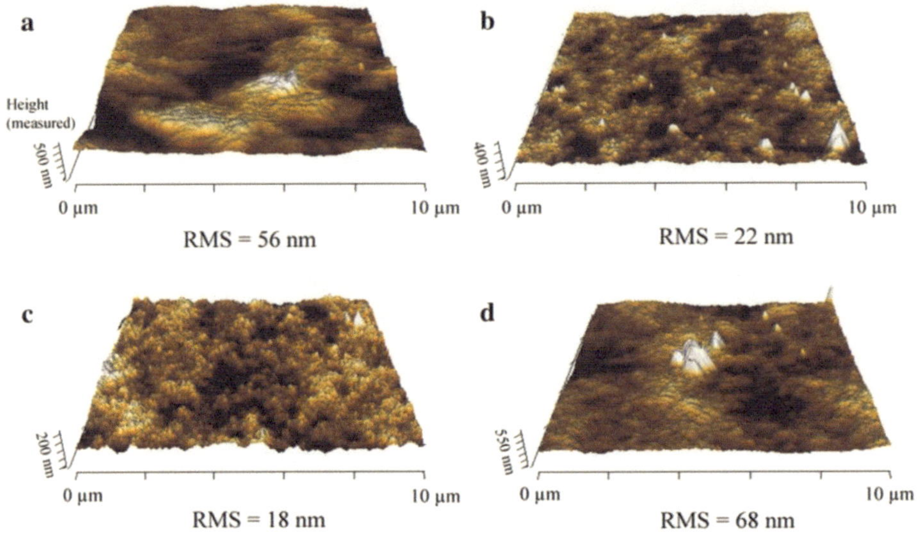

Figure 25.
AFM images of the NBC film surfaces are shown as a three-dimensional structure with their root mean square (RMS) roughness value: a cellulose film; b (2%), c (5), and d (15%) cellulose film containing 2, 5, and 15% nanochitosan particles, respectively [18].

be very challenging [70]. Niroomand et al. reported the use of AFM to monitor the effect of nanochitosan addition on cellulose NBCs As can be seen in **Figures 25** and **26**. The 3-D AFM overviews showed a decrease in transparency and roughness of the cellulose-based NBCs when dosages of nanochitosan were increased, compared to the addition of 15% nanochitosan, which resulted in a rougher surface [18].

3.3 Hydrophilicity of nanobiocomposite

Hydrophilicity is highly dependent on the NBC morphology, surface chemistry, water absorption, solid state, and porousity of an NBC. It greatly affects the wettability and dispersibility of NBCs and adhesion to the mucous membranes, and ability to adjust to curves and uneven textures like skin [18]. It is usually determined by monitoring the contact angle (CA)of a water droplet on a surface on NBCs film or matrix using a goniometer or a contact angle analyzer. The spreading out of a water droplet on an NBC surface indicates its hydrophilicity, while the resistance to spreading indicates hydrophobicity [77]. A low CA (below 90) between the droplet and the surface indicates hydrophilicity and wettability [18].

It has been reported that polysaccharides like chitosan and alginate are highly wettable; however, they have low mechanical strength. Their ability to absorb water is due to the presence of hydrophilic groups on the surface of their relevant NBCS. Thus, NBCs of these polymers usually have a hydrophilic nature. Once functionalized, the polymer will change the NBCs wettability depending on the functional group exposed to the surface. Additionally, the presence of a more compact microstructure of the nanocomposites due to the strong interaction within a polymer or the NBC will affect its wettability. It is important that surface hydrophilic group are free to interact with water in order to enhance wettability and is not fully engaged. Moreover, the presence of an amorphous state in an NBC allows the free movability of hydrophilic groups to interact with water, hence enhancing its water absorption and wettability [18].

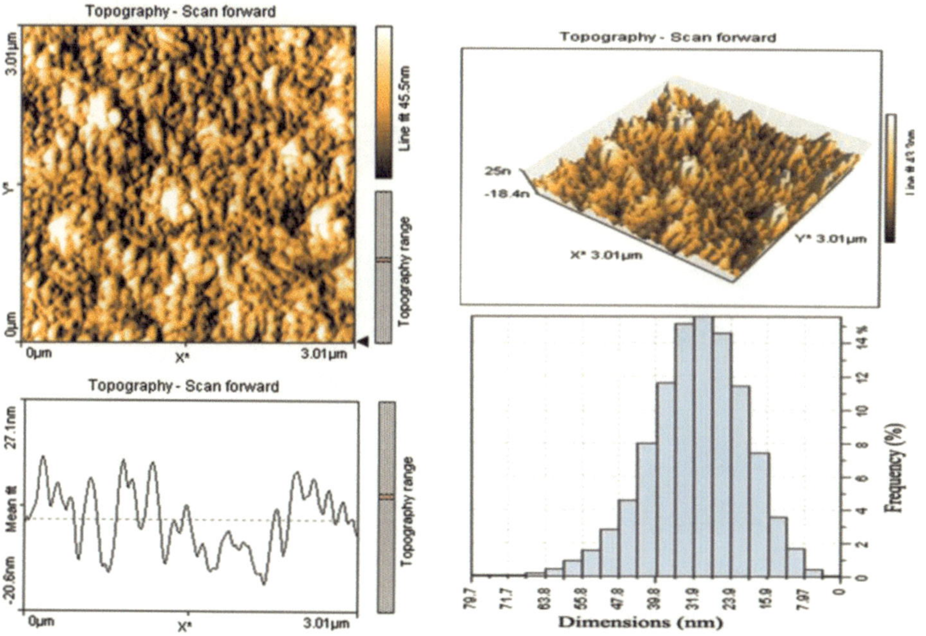

Figure 26.
AFM image of the nanochitosan particles and the histogram demonstrating the average size and size distribution of the nanochitosan particles [18].

Espino-Pérez et al. reported the increased hydrophilicity on NBCs based on cellulose nanocrystals (CNC) surface following esterification with phenylacetic acid and hydrocinnamic acid [52]. Similarly, Chiaoprakobkij et al. evaluated the water contact angle of films based on mechanically-disintegrated bacterial cellulose, alginate, and gelatin (BCAGG), plasticized with glycerol before and after curcumin loading (BCAGG-C). The study showed an increase in the angle with increasing curcumin content, as shown in **Figure 27** 4A–D. The water contact angle of the drug-free NBCs was 49.5, while after loading with curcumin concentrations (2, 4, 6, and 8 mg/mL), the contact angle ranged 54.7–73.3, indicating the hydrophilic nature of the drug-free NBCs [78].

3.4 Particle size distribution and determination and surface charge

NPs size plays an important role in cellular uptake and fate of the NPs within the body, consequently it affects the drug's half-life and therapeutic efficacy. In addition, the particle charge (zeta-potential) has a large impact on surface recognition, surface interaction with biomolecules, and cellular targeting. Importantly, size and zeta-potential contribute significantly to the NP storage stability [79, 80]. The need for a monodisperse NP dispersion mandate the use of very advanced and facile methods that facilitate monitoring of NP size over time and following exposure to changes in temperature and pHs. The most recent methods in NPs size determination are dynamic light scattering (DLS) and laser Doppler electrophoresis (LDE) [80, 81].

Visually, colloidal samples should be slightly hazy with a homogenous appearance termed the Tyndall effect. **Figure 28** depicts NPs based on chitosan succinate 0.1%

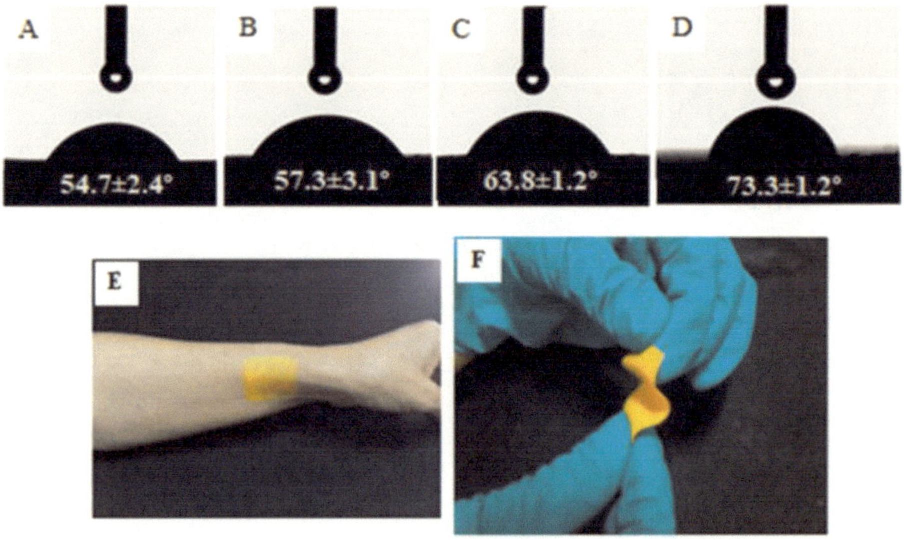

Figure 27.
Water contact angles of NBCs based on bacterial cellulose, alginate, gelatin: (A) 2 mg/mL; (B), 4 mg/mL; (C), 6 mg/mL; and, 8 mg/mL (D)), solutions of curcumin at concentrations of 2, 4, 6 and 8 mg/mL [78].

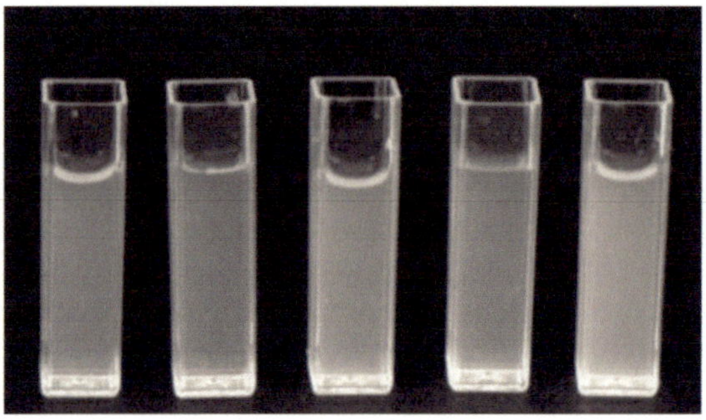

Figure 28.
Responses of ionotropic chitosan succinyl amide NPs (CS-H-80-TPP/EDC) to variable $CaCl_2$ concentrations [7].

NPs crosslinked with polyphosphate and dispersed in 4.8 mM HCl [75]. In a colloid, the light scatters in different directions due to the colliding dispersed particles. It occurs when the diameter of an NP is in the range of roughly 40 to 900 nm, i.e., somewhat below or near the wavelengths of visible light (400–750 nm). Light scattering will occur when the diameter of dispersed particles is much smaller than the wavelength of light used [7, 75].

DLS technique is based on exposing NP dispersion to a monochromatic coherent laser beam where the particle will move as a result of Brownian motion. With continuous collision between the particles, the distance between them will also change and hence fluctuations of the phase relations of the scattered light will be detected [82, 83]. Since the particle has different diameters, the number of particles within the scattering volume will vary in sedimentation with time. By observing the change in light intensity, this will be digitally correlated by photon analysis [84]. The DLS system measures the rate of intensity fluctuations and then uses this to calculate the size of the particles as defined in the Stokes-Einstein equation (Eq. (3) [81–82]:

$$D = \frac{\bar{R}T}{NA} \frac{1}{6\pi\eta\, r} \tag{3}$$

where, D is the diffusion constant; $\bar{R}$ is the gas constant; T is the absolute temperature; η is the dynamic viscosity; r is the radius of the spherical particle; NA is Avogadro's Number. Depending on the dispersity of the sample particle, light intensity and its decay will be correlated to give an idea about the size distribution of the sample NPs. The quality of measurement depends on the sample and the measuring device (the laser source, the detector precision, the correlator software, etc.) [80, 81].

It should be emphasized that several factors contribute to the accuracy of measurements. For example, a high sample concentration will affect the length of the path, allowing for the particle collisions resulting in a small path; hence, multiple collisions interfere with the measurement. Simple DLS instruments can measure at a fixed angle to determine the mean particle size in a limited size range. Since NBCs are mostly characterized by polydispersity, the light scattering by large particles can overwhelm the one by smaller particles. A multi-angle instrument that allows full particle size distribution is needed for accurate readings [80, 81].

In order to measure the size using this technique, the sample should be dispersed in the correct solvent (known refractive index, RI) at the optimum concentration and proper dilution. DLS measurement for dry samples can be achieved following dispersion in the proper solvent, as described before. The most commonly reported dispersion media is double deionized water (RI 193.39 nm at 25C) for pharmaceutical applications. In addition, glycerol and ethanol were also used with diluted concentrations [81]. Other measurements were done using methanol and toluene. The stability of the sample in the solvent is crucial, i.e. the particles should remain dispensed (undissolved with no aggregation) during the measurement. The purity of the dispersed particle and solvent and the exclusion of fiber and dust are highly needed prior to applying this technique appropriately [81, 82]. **Figure 29** depicts an example of the particle size distribution of DLS investigation with NP dimensional distribution: NP from 0.5% PNIPAM/PVA concentration with various methyl oleate concentrations (top); nanoparticles from 5% PNIPAM/PVA concentration with various methyl oleate concentrations (bottom) [29].

Published data related to NBC size characterization include but are not limited to grafted pectin [13], chitosan phthalate and phenyl succinate [7, 79], thiolated chitosan [5], chitosan [30, 85], cellulose [86] and many others.

3.4.1 NBC surface charge (Zeta potential) measurement

The zeta potential is a measure of the electric charge at the surface of NPs, being an indirect assessment of their physical stability. The surface properties of NBCs will greatly impact the affecting the release properties and the interaction between the drug delivery system and the cellular receptor. The presence of specific chemical groups on the surface of the NBCs will also enable further functionalization by antibodies or by enzymes [41]. These properties greatly affect drug targeting resulting in site-specific drug delivery, for example, in tumor treatment [46].

Laser Doppler electrophoresis (LDE) involves measuring the change in the light scattering intensity due to the shift in the frequency of the wave as a result of interaction between a particle surface charge and the electric field. The direction and velocity of the motion are a function of particle charge, the suspending medium, and

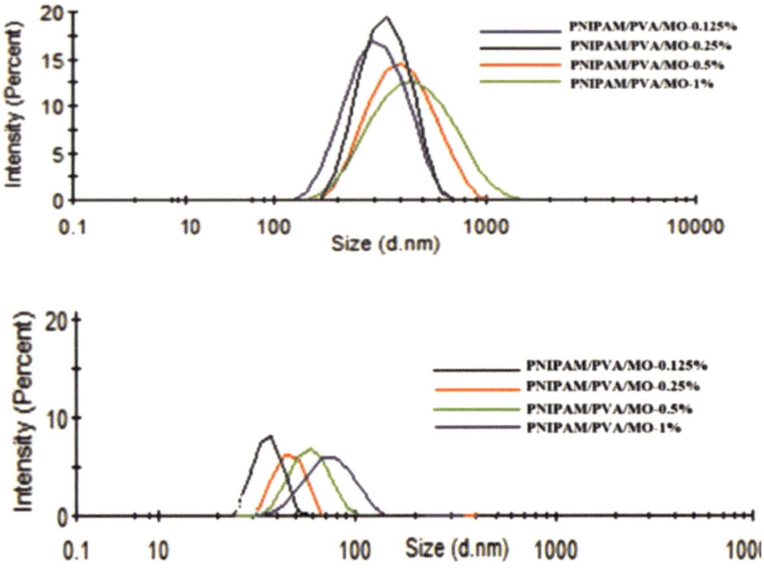

Figure 29.
DLS investigation with nanoparticles' dimensional distribution: nanoparticles from 0.5% PNIPAM/PVA concentration with various methyl oleate concentrations (top); nanoparticles from 5% PNIPAM/PVA concentration with various methyl oleate concentrations (bottom) [29].

the electric field strength. Particle velocity is then measured by observing the Doppler shift in the scattered light. The particle velocity is proportional to the electrical potential of the particle at the shear plane, which is zeta potential [80].

When NPs are dispersed in water, they will gain electric charges. As a consequence, a concentration of oppositely charged ions (counterions) builds up at the particle surface. If these counterions are separated from or sheared off the particle by electrophoresis, a streaming potential can be measured in mV (**Figure 30B**). The measurement usually involves converting the electrophoretic mobility by the equipment software into zeta potential data through Smoluchwski's approximation [24]. When performing the zeta potential measurement, it is important that the pH and the temperature are controlled with continuous dispersion of the NBC [83]. A negative

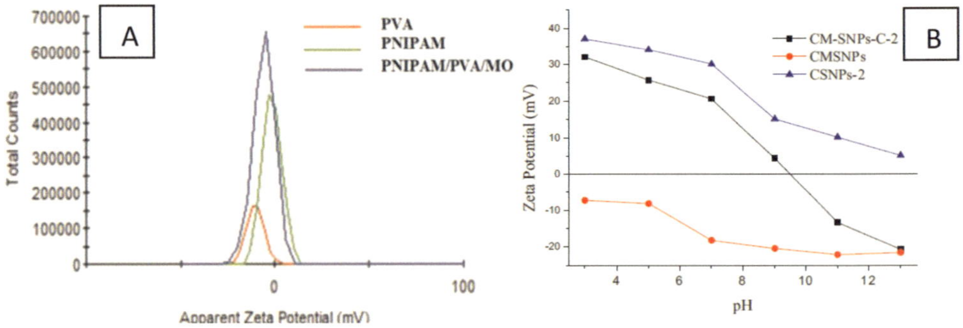

Figure 30.
(A) ZETA potential of PVA, PNIPAM, and PNIPAM/PVA/MO nanoparticles (a); Schematic representation of nanoparticles core-shell structure (b) [29]. (B) Effect of initial solution pH on zeta potential of amphoteric SNPs (CM-SNPs- C-2), CMSNPs, and CSNPs-2 [56].

charge in an NBC surface favors the stability and prevents aggregation (self-accumulation) of NPs. Zhou et al. reported enhanced stability of negatively charged sinapic acid-grafted-chitosan NPs loaded with black rice anthocyanins when compared to chitosan NPs alone [39]. Wang et al. reported the presence of a negative surface charge following grafting succinyl chitosan, which can be attributed to the ionizable –COOH group present at the NP surface at pH 7.4 [41]. Similar observations were reported by Chang et al. while monitoring the zeta potential change with pH in starch-modified NPs [56] (**Figure 30B**).

Many published data have reported the use of zeta potential in NBCs characterization. The following are examples: cellulose NBCs [86], glucose-conjugated chitosan [48], starch [44], pectin [43], chitosan [7, 46], lecithin/chitosan phthalate and Phenyl succinate [79], etc.

3.5 Adsorptive properties

Surface adsorptive properties of NBCs, have a significant impact on its application as a drug delivery system, for example, for surface drug loading, functionalization by ligands (antigens, antibodies, etc.), enzyme immobilization, etc. [87, 88]. Additionally, It affects NBCs recognition, phagocytosis, elimination by macrophages, which can then further affect their transport and fate in the body [3, 15, 89].

Depending on the type of biocomposite, adsorption is highly affected by pH, ionic strength, adsorbent concentration, contact time, and temperature. Adsorption experiments can usually be conducted by shaking the NBCs or the polymer with variable concentrations of an adsorbent in a suitable container for a proposed time interval. The solutions are usually agitated at a constant speed in a temperature-controlled water bath at different temperatures for the required period. At a predetermined interval, samples are withdrawn, centrifuged, and the concentration of the adsorbent is analyzed using a suitable quantifying technique like UV [90, 91].

Natural polymers like chitosan have $–NH_2$ and –OH adjacent to $–NH_2$ in its backbone, which enhances its adsorption properties for many metal ions such as aluminum, silver, zinc, etc. [90]. Lee et al. investigated the adsorption of mucin to chitosan (mucoadhesiveness) during evaluating thiolated CS Intranasal delivery of theophylline [92] Similarly, the hydroxyl groups on the surface of nanocellulose(cellulose nanofibrils and cellulose nanocrystals) allowed electrostatic adsorption, making them suitable for enzyme/protein immobilization [93].

Adsorption evaluation generally involves gentle mixing of a precise volume of NBCs dispersion with a fixed volume of drug, enzyme, or metal solution, etc., with continuous shaking at a specific speed in a thermostat-regulated shaker at 25°C. Once equilibration is attained (e.g., 20 h), the mixture is filtered, and the concentration of residual adsorbent (enzyme, etc.) in the filtrate is determined by UV–VIS spectrophotometry. Oshima et al. reported this method in measuring the adsorption of protein (lysozyme) to the surface of phosphorylated cellulose using Eq. (4) [91]:

$$\%Adsorption = \frac{Co - Ce}{Co} \times 100\% \tag{4}$$

$$Amountadsorbed(q)[umol/gm] = \frac{Co - Ce}{W} \times V$$

Where C_0 and C_e are the protein concentrations before and after adsorption in mol/ml, W is the dry mass of adsorbent in gm, and V is the volume of solution in ml.

Adsorption isotherms of enzymes like lysozyme can be obtained at constant temperature (for example, 30°C) using aqueous solutions containing varying concentrations of lysozyme and adsorbents [91].

The effect of molecular weight on NBCs has a great role in adsorption properties. Riegger et al. investigated the impact of molecular weight of six commercially available, highly deacetylated chitosan based NP on the adsorption of diclofenac and carbamazepine. They reported an adsorption capacity of up to 351.8 mg/g diclofenac for low MW chitosan NPs, and all chitosan NPs showed superior adsorptions when compared to untreated chitosan. Hence, the results suggested the use of the prepared chitosan NPs as promising adsorbers for diclofenac and carbamazepine [85].

3.6 Gelling properties

Gel formation can be obtained using many natural polymers like alginate and chitosan, etc., which are capable of forming a 3Dimetnional structure upon crosslinking of their polymeric chains. Remarkably, a hydrogel is one type of gel that swells upon exposure to water, enabling its use in the delivery of many drugs [94]. The type and degree of crosslinking affect the nature of the gel, either tough or soft gel, the solubility, and its mechanical strength [4, 95]. There are two types of methods to prepare gels: physically (by a change in pH or temperature) and chemically (by electrostatic, covalent crosslinking, biological cell crosslinking, free radical polymerization, and click chemistry) [40, 66]. In general, ionic gelation favors mild conditions and results in soft gels (solvent, pH and temperate, etc.) [96]. Alginate has dominated among all hydrogels and is the most widely used hydrogel for encapsulation due to its low cost, high availability, and durability, as well as its nontoxicity to host organisms and well-established encapsulation process. When multivalent cations like Ca^{2+} are present in an aqueous solution, certain polymers like alginates and the like have the necessary characteristics to construct suitable matrices [11, 97]. Alginate has been successfully used to encapsulate cells [98].

On the other hand, covalent crosslinking involves harsh conditions (toxic materials like glutaraldehyde, high temperature, etc.) and favors more tough gels [29, 66]. Biocomposites like pectin/chitosan gel prepared by the casting method have been optimized by varying their components using lactic acid or glycerol as solvents. Also, an antibacterial test against *Bacillus subtilis* confirms that the pectin and chitosan retains their antibacterial property in biocomposite materials [99].

3.6.1 Visual, optical transparency (clarity) and surface evaluation of hydrogels

The clarity and surface smoothness of gels mainly depends on the presence and structure of insoluble components [100]. The transparency of a gel indicates the solubility of the components and homogeneity of the fabrication process. Direct evaluation of visual transparency is performed by the eye, while the optical transparency of the formulated hydrogel is analyzed by a UV as described previously [101]. 100% light transmittance in distilled water indicates the optical transparency of a hydrogel. It is important that both of these measurements are performed at different temperatures, for example, 25 and 37°C, accompanied by pH measurement. Additionally, SEM analysis is usually performed to evaluate the smoothness, homogeneity, or heterogeneity of the gel surface [66].

3.6.2 Sol–gel transition behavior, gelation time, and gelation temperature

A sol–gel transition occurs through additional intermolecular interactions of a hydrophobic nature, leading to the formation of a turbid gel that can be achieved by exposure to variable temperature, pH, or shaking. It must be determined at a physiological temperature 37°C) since it will greatly affect the drug release, injectability and storage conditions [96]. The gelation temperature is usually measured by placing the polymer solution in a glass vial and exposing it to heat with gentle shaking. The content is observed for gelation on intervals while inverting the glass vial at a 90° angle for 1 min. Once flowability is stopped, the temperature is recorded as a gelation temperature. The time needed for a solution to stop flowing (gel formation) is termed gelation time [96, 102].

3.6.3 Sol–gel fraction

Sol–gel fraction examination is carried out to determine the sol and gel fraction in any prepared hydrogel. The liquid portion of the hydrogel is expressed as the sol fraction. In this test, discs of the dried hydrogel are weighed(W_i) and kept in boiling water at 100°C for approximately 4 h. After a certain time, the discs are removed from the water bath and dried at room temperature for 24 h or a low-temperature oven to a constant weight (W_d). For calculating the sol and gel fraction of the hydrogel, Eq. (5) is employed [103]:

$$GF\% = \frac{Wd}{Wi} \times 100\% \tag{5}$$

$$\mathrm{s} = 1 - \mathrm{GF}$$

where s is the sol fraction, GF is the gel fraction.

3.6.4 Water absorption capacity (WAC)/fluid uptake ability

This test usually involves immersing an accurately weighed film or hydrogel in water or suitable fluid at room temperature and allowing it to equilibrate. Following a specific time (6 and 24 h), the sample is removed, and the surface water fluid is removed (wiped) gently and re-weighed. Water content /fluid uptake (%WAS) is usually determined by a precise balance and calculated using the Eq. (6) [103]:

$$\%WAC = \frac{Ws - Wd}{Wd} \times 100\% \tag{6}$$

where Wd and Ws are the weights of the dry sample and wet sample, respectively. A similar procedure is applied in measuring the Fluid uptake ability by immersing the weighed samples in PBS (pH 7.4) and artificial saliva (pH 6.2) at 37°C.

Additional tests to characterize NBCs-based gels include:

3.6.5 Surface wettability

This can be evaluated as a static water contact angle by monitoring a water droplet from different locations using a contact angle analyzer [103].

3.6.6 Hydrogel oxygen and water permeability

The oxygen transmission rate (OTR) of the films is needed for dermal dressing to ensure the non-occlusiveness of hydrogel/films. Usually, it is determined by an oxygen permeation analyzer at 25°C and 0% relative humidity. In this test, one side of the sample is exposed to a nitrogen atmosphere, while the other side is exposed to an oxygen atmosphere. When the concentration of oxygen on the nitrogen side becomes constant, the test is considered complete [104].

On the other hand, water vapor permeation is measured by a water permeability analyzer. The pre-weighed sample is placed in a test dish containing a desiccant, and the assembly is placed in a controlled atmosphere at 37°C and 98% relative humidity. Periodic weighting is performed to determine the rate of water vapor movement through the specimen into the desiccant and plotted against time [103].

3.7 Mechanical properties

The need to measure the mechanical properties (elasticity and flexibility) is mandatory in formulations applied directly to the skin or the tissues. NBCs hydrogels are flexible, porous and can be fabricated by chemical or physical crosslinking nanomaterials as described previously. Varying the conditions of the crosslinking process (crosslinker type, time, temperature, etc.) can be used to achieve a strong, flexible hydrogel [105].

Among the important mechanical properties are tensile strength (TS) and elongation at break (EB), which are measured using a tensile strength tester. TS and EB are usually calculated using the Eq. (7) [106]:

$$\text{TS} = \text{Maximum force} : (\text{Film thickness}) \times (\text{Film width}) \tag{7}$$

$$EB = \frac{\Delta L \times L0}{100}$$

Where ΔL and L0 are the elongation of the specimen at the moment of break, and the initial length of the specimen, respectively.

The effect of additives on the mechanical strength of NBCs has been studied by Kassab et al. They investigated their mechanical reinforcement capability for k-carrageenan biopolymer on cellulose nanocrystals (CNC). The obtained CNC was dispersed into a k-carrageenan biopolymer matrix at various CNC contents (1, 3, 5, and 8 wt%), and the prepared films were further characterized. The incorporation of CNC enhanced the mechanical properties compared with the neat k-carrageenan (k-CA) film, as seen in **Figure 31**. All nanocomposite films have higher tensile strength compared to films based on neat k-CA biopolymer. This is attributed to the great improvement attained by the addition of CNC. Furthermore, the researchers reported an increase in the modulus and strength by increasing the CNC content from 1 to 8 wt %, with slight variation in the toughness of CNC of the biocomposite [65].

Chaichi et al. implemented a statistical optimization approach to study the effect of additives like Ca^{2+} as a crosslinker and glycerol on the tensile strength of NBCs. The researchers demonstrated that Ca^{2+} ions could significantly reduce the swelling and elongation to break while increasing the tensile strength of the NBCs [106]. In a similar study, Chiaoprakobkij et al. reported curcumin-loaded film's formulation based on bacterial cellulose/alginate/gelatin using mechanical and casting methods.

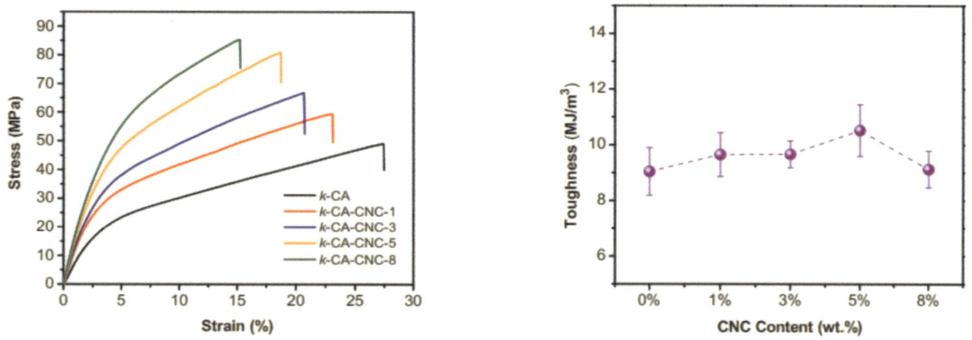

Figure 31.
Typical stress–strain curves of neat k-carrageenan (k-CA) film and its nanocomposites at different CNC contents (1–8 wt%) [65].

Films were stretchable with the appropriate stiffness and enduring deformation, which enabled dermal application when sufficiently hydrated. Additionally, the films have good mucoadhesive properties, which enhance the antibacterial activity of curcumin against *E. coli* and *S. aureus* [78].

3.7.1 Tensile strength and elongation to break

The consistency among the polymer chains, flexibility (elongation before breakage), and ability to resist extension can be measured using tensile strength [106]. Adam et al. studied the effect of incorporating gum Arabic κ-carrageenan biocomposite in hydroxypropyl methylcellulose (HPMC) hard capsules shell. The optimization involved the use of variable ratios of the hard capsule constituents in order to achieve a capsule with good tensile strength and optimum disintegration time. The researchers suggested that this biocomposite can be an alternative to ordinary gelatin used in capsule shell formulation [107].

3.7.2 Stiffness of the material

Kurowiak et al. described the measurement of sodium alginate-based hydrogel when subjected to static tensile testing to determine its elasticity (Young's modulus). The studies showed that hydrogel crosslinked with calcium ions showed a lower mechanical strength compared to the one crosslinked with Ba^{2+} cations (**Figure 32**). The researchers attributed the difference to the increased barium affinity to alginate monomers (G blocks), resulting in a hydrogel by forming the egg-box structure, characteristic of alginate NPs [108].

3.8 Rheological properties

Rheology is the study of how materials deform when a force (shear) is applied to them. Rheological properties affect fabrication conditions and the quality of the fabricated products. These materials are mostly liquids or liquid-like materials. Rheological measurements are also very useful for characterizing the flow properties of emulsion systems and predicting their behavior during manufacturing, storage, and drug administration [66, 109].

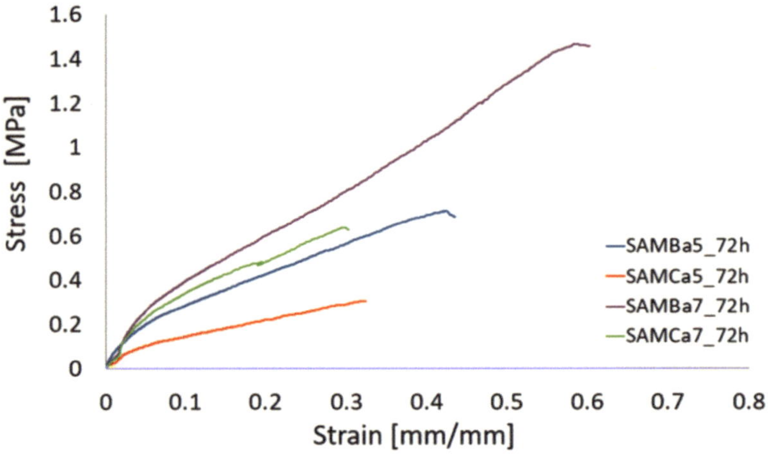

Figure 32.
Strain–strain analysis plot for the proposed sodium alginate-based material cross-linked with Ca^{2+} or Ba^{2+} cations using different cross-linking times after 72 h cross-linking time [108].

Basically, the rheological properties of NBCs are affected by the natural polymer properties and the additives included within the formula and the technique used in the formulation. The control of these two factors allows the optimization of the formula to suit the application in relation to the route of administration, for example, ocular, nasal, at the tumor site by injection, etc. Properties that can affect the rheology of NBCs include molecular weight (MW) and its distribution (MWD), morphological, molecular structure, and orientation under electric or magnetic fields. High molecular weight polymer exhibits thixotropic behavior with high resistance to extreme temperatures, freeze–thaw cycles, pHs, and salt concentrations [15]. Formulation factors include the presence of an additional compound or impurities (crosslinkers, surfactants, stabilizer, etc.), the solvent used, ionic strength, pH, NBCs concentration, pressure, and temperature [71]. It should be noted that in strongly crosslinked samples, no rheological measurements could be performed due to their brittle properties. Sample assessment includes carful control of temperature and prevention of the solvent evaporation [66, 71].

Rheometers can be divided into two categories: rotational and capillary types. Two major types of rheological experiments can be performed utilizing parallel-plate or rotational rheometer, the sweep tests (varying strain, frequency, and temperature) and the steady shear sweeps (**Figure 33**) [110]. The principle of each test and examples are described in the following paragraphs.

3.8.1 Flow curves (steady shear flow)

The importance of having a consistent viscosity during storage is a vital feature of drug delivery systems. Flow curves describe the rheological behavior of a material, more specifically, the dependency of the viscosity on the applied shear rate and the tendency of a material to flow. The plot is represented by viscosity as a function of shear rate (log relationship can be used). These are usually used to evaluate the viscosity of hydrogels using different crosslinking ratios. Formulation of NBCs in the nanoscale can increase the viscosity. Ahmad et al. reported that the viscosity of starch-based NPs dispersion was influenced by the shape, size, and distribution of the starch

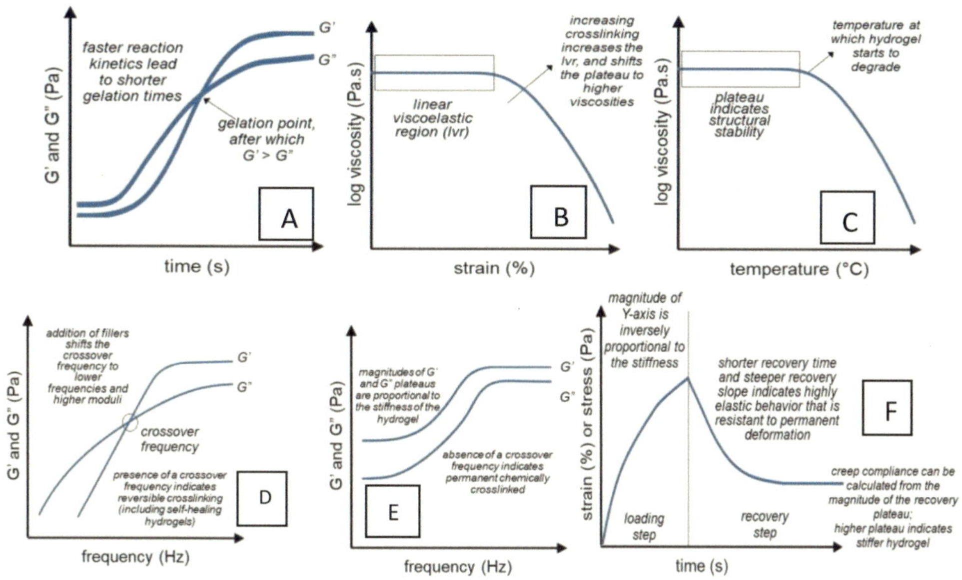

Figure 33.
(A)Time sweep, (B) Strain sweep, (C)Temperature sweep, (D) Frequency sweep, and (E)Creep Compliance, (F)Creep Recovery [95].

granules and also by the amylose content with variable viscosity depending on the source of starch. As can be shown in **Figure 34**, a decrease in the size of starch at the nanoscale increased the viscosity of the dispersion compared to the native starch [44].

Mishra et al. reported improved shear stability of the polyacrylamide-grafted pectin hydrogel compared to the pectin-based hydrogel. The viscosity of the polymer solutions decreases with an increase in shear rate. Both the aqueous 5% solutions of grafted Pectin and Pectin showed strong pseudoplastic behavior. As can be seen in

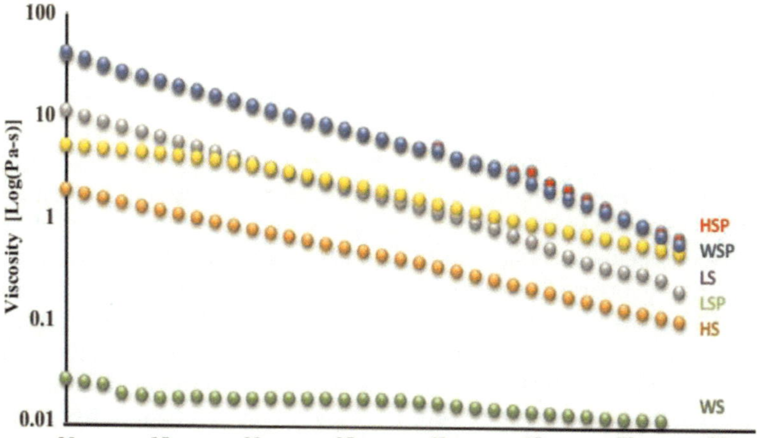

Figure 34.
Flow curves (steady shear flow) for native and nano starch particles against shear rate.: Horse chestnut particle (HSP), Water chestnut Particle (WSP), Lotus stem particles (LSP), Horse chestnut (HS), Water chestnut (WS), and Lotus stem (LS) [44].

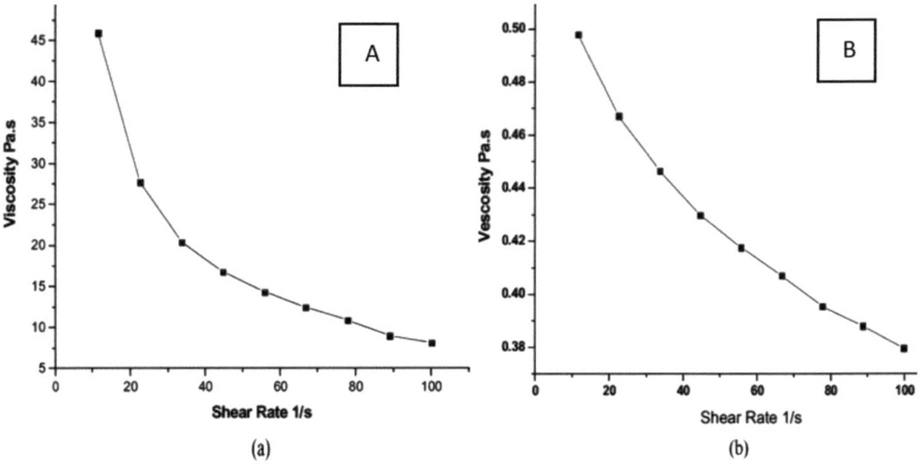

Figure 35.
Viscosity versus shear rate curve of 5% grafted pectin solution(A); Viscosity versus shear rate curve of 5% pectin solution(B) [49].

Figure 35, at low and high shear rates, the viscosity of the grafted pectin solution was higher than the pectin solution. This suggests that grafted pectin solution was more shear stable than the ungrafted pectin, which can be attributed to longer branches in the grafted pectin [49].

In addition, studying the viscosity dependence on the shear rate enables the classification of the hydrogel into those which exhibit thixotropy or its opposite phenomenon, rheopexy behavior. In this test, the sample is exposed to an increased shear rate, and the viscosity of the hydrogel decreases up to a certain minimum which indicates thixotropic behavior. After this, the shear rate is reduced, which leads to increased values in the viscosity, which are higher than the original viscosity values for the respective shear rate. This phenomenon is known as negative thixotropy or rheopexy.

3.8.2 Time sweep test

This test evaluates structural changes for a specific material after applying a shear over a certain time. These changes can be observed following evaporation of the solvent, curing, gelation, polymer degradation, or recovery. For example, the gelation time can be related to the kinetics of the gelation reaction, which is defined as the crossover point of the storage (G′) and loss (G″) modulus [95]. It should be emphasized that no rheological measurements could be performed for strongly crosslinked samples due to their brittle properties [111]. Stanescu et al. reported that the uncrosslinked samples of bacterial cellulose (BC)/chitosan NBCs loaded with silver sulfadiazine showed the lowest shear viscosity values compared to crosslinked NBCS (**Figure 36A**) which, can be attributed to network development during crosslinking process. The prolonged exposure to the crosslinker resulted in a higher shear viscosity. Additionally, the presence of BC reduces CS shear viscosity when compared to CS alone (**Figure 36B**), which showed higher shear viscosity values [29].

Strain sweep test (amplitude sweep).

This test is used to characterize hydrogels using increasing oscillatory strain at a constant frequency on the storage (G′) and loss (G″) modulus of the hydrogel to determine the linear viscoelastic region (LVR) (**Figure 37A**) Jannatamani et al.

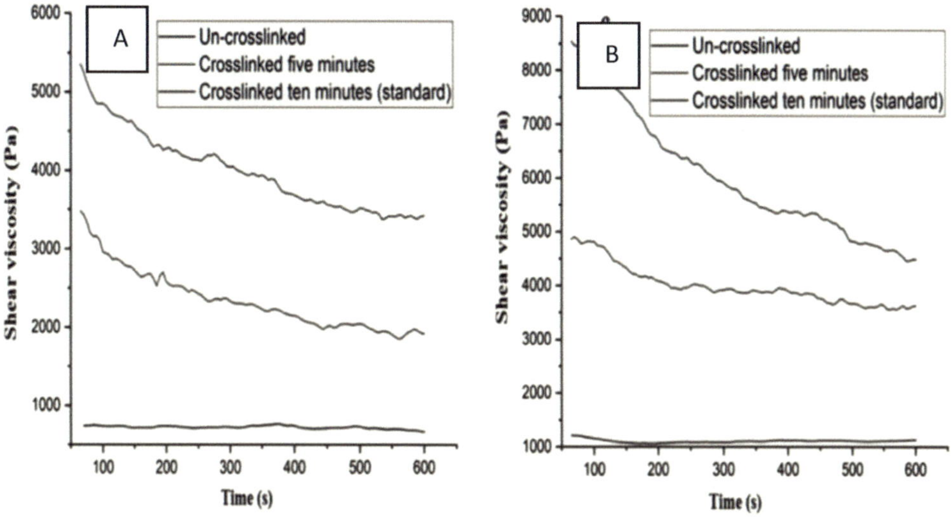

Figure 36.
Rheological behavior (shear viscosity) measurements for (A) Bacterial Cellulose (BC)/chitosan and (B) chitosan alone [29].

evaluated nano-hydrogels and films based on the wood Cellulose NanoFibers (WCNF), Bacterial Cellulose NanoFibers (BCNF), and Chitin NanoFibers (ChNF LVR). At low shear stress, the moduli are independent of the increasing stress. However, as the stress is increased, the G′-G″ crossover point potentially reaches, at which the gel–sol transformation occurs, and the material starts to behave like a fluid. Additionally, a Strain sweep in hydrogels can be used to estimate the threshold strain required above which shear thinning behavior is observed (**Figure 37B**) [112].

3.8.3 Temperature sweep

Sometimes it is termed the temperature ramp test, which enables predicting the structure of the hydrogel, and its stability when subjected to a certain range of

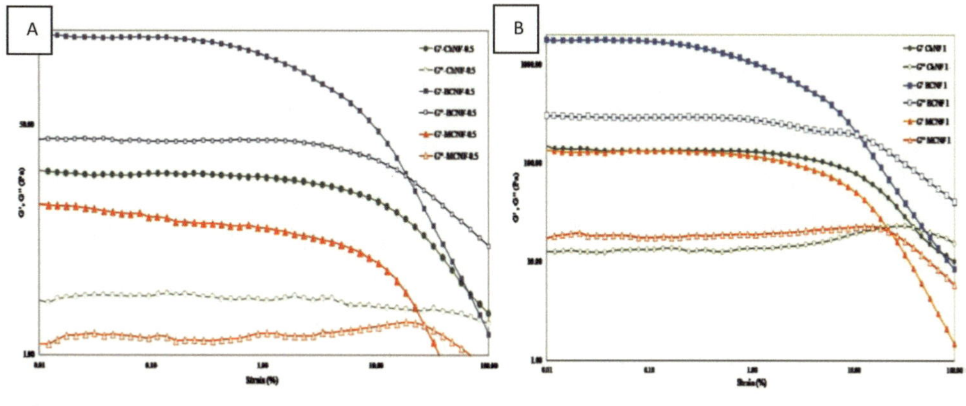

Figure 37.
Storage and loss modulus of 0.5 wt% (A) and 1 wt% (B) concentration of WCNF, BCNF, and ChNF nano-hydrogels as a function of strain [112].

temperatures. This can be achieved by studying the storage (G′) and loss (G″) modulus in a certain temperature range with an evaluation of the sol–gel transition of the hydrogel (**Figure 33C**). The point at which the viscosity drops suggests the temperature above which the hydrogel starts to degrade or de-crosslink. Plotting the viscosity vs. temperature will enable predicting the temperature when a hydrogel will degrade or uncrosslinked [95].

3.8.4 Frequency sweep

The effect of additives on viscoelastic properties of hydrogel can be studied by varying the frequency and evaluating its relationship with the storage (G′) and loss (G″) modulus (**Figure 38**). Ajovalasit et al. used a frequency sweep test to evaluate the impact of additives like glutaraldehyde glycerol and PVA on the properties of Xyloglucan-based hydrogel films for wound dressing. They found that the addition of glycerol does not impact the rheological properties, whereas the addition of glutaraldehyde moves the G′ and G″ crossover point to lower frequencies. Interestingly, the addition of PVA decreases the storage and loss modulus values (G′ and G″) [113].

3.8.5 Creep compliance, creep recovery, and stress relaxation

This test is used to evaluate the elasticity of hydrogel films when a sample is subjected to a constant static load (strain) and how the structure recover following withdrawing this strain. In addition, it enables predicting hydrogel behavior following frequent use in real. Thus an increasing strain reaches an equilibrium after a certain time. After that, no further stress is applied, and the recovery of the sample is recorded over a certain fixed time [95].

Stress relaxation is the inverse of the creep compliance test, where a stress relaxation test subjects the sample to a constant strain and measures the stress exerted by the sample. It gives an idea of how well materials can dissipate stress over time at a constant strain. Craciun et al. evaluated the rheological properties of chitosan-based hydrogel compared to chitosan/ pyridoxal 5-phosphate (vitamin B6 precursor) based hydrogels used for local action, in tumors or on wounds. Chitosan (NH_2 source) and

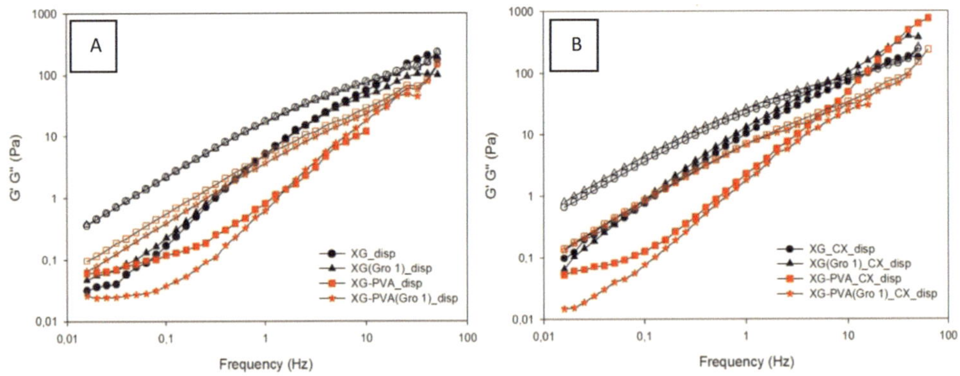

Figure 38.
Dynamic mechanical analysis in frequency sweep mode of precursor dispersions and films. Solid symbols: storage modulus; open symbols: loss modulus. (A) Aqueous XG and XG/ PVA before (XG_disp and XG-PVA_disp) and after addition of glycerol (XG(Gro 1)_disp and XG-PVA(Gro 1)_disp). (B) Precursor aqueous dispersions of the chemical films obtained by addition of glutaraldehyde in the same samples illustrated in panel (a) [113].

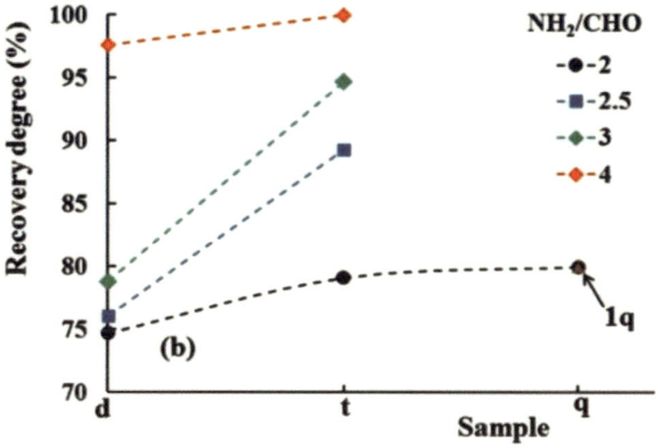

Figure 39.
Effect of NH_2/aldehyde ratio on the structure recovery ability determined by the continuous step strain measurements with various dilutions and NH_2/CHO ratios [40].

the aldehyde of pyridoxal (CHO) form hydrogel simultaneously due to imine group formation. The study revealed the dependence of gel formation on the ratio of NH_2/ aldehyde ratio of chitosan /pyridoxal 5-phosphate (**Figure 39**). A higher recovery degree was achieved when the ratio was less than 3 [40].

4. Conclusion

Nanobiocomposites have been investigated in many fields, including the medical and pharmaceutical fields. For a successful application of NBCs as a drug delivery system, it is essential to perform an extensive physicochemical characterization of these NBCs. Each property discussed in this chapter has an extreme effect on the final appropriateness of these NBC as a drug delivery system. The impact of these properties can vary from drug-polymer interaction to therapeutic efficacy, safety, and stability. Therefore, a comprehensive characterization of each property will enhance achieving a safe and effective drug delivery system.

Author details

Isra Dmour
Faculty of Pharmaceutical Sciences, Department of Pharmaceutics and Pharmaceutical Technology, The Hashemite University, Zarqa, Jordan

*Address all correspondence to: isradmdm@gmail.com

References

[1] Dmour I, Taha MO. Natural and semisynthetic polymers in pharmaceutical nanotechnology. In: Organic Materials as Smart Nanocarriers for Drug Delivery. New York, USA: William Andrew Publishing; 2018. pp. 35-100. DOI:10.1016/B978-0-12-813663-8.00002-6

[2] Basim P, Gorityala S, Kurakula M. Advances in functionalized hybrid biopolymer augmented lipid-based systems: A spotlight on their role in design of gastro retentive delivery systems. Archives of Gastroenterology Research. 2021;**2**(1):35-47

[3] Mikušová V, Mikuš P. Advances in Chitosan-based nanoparticles for drug delivery. International Journal of Molecular Sciences. 2021;**22**(17):9652. DOI: 10.3390/ijms22179652

[4] Huo Y, Liu Y, Xia M, Du H, Lin Z, Li B, et al. Nanocellulose-based composite materials used in drug delivery systems. Polymers. 2022;**14**(13): 2648. DOI: 10.3390/polym14132648

[5] Ouerghemmi S, Dimassi S, Tabary N, Leclercq L, Degoutin S, Chai F, et al. Synthesis and characterization of polyampholytic aryl-sulfonated chitosans and their in vitro anticoagulant activity. Carbohydrate Polymers. 2018; **196**:8-17. DOI: 10.1016/j.carbpol.2018.05.025 Epub 2018 May 8

[6] Almukhtar JGJ, Karam FF. Preparation characterization and application of Chitosan nanoparticles as drug carrier. Journal of Physics Conference Series. 2020;**1664**:012071. DOI: 10.1088/1742-6596/1664/1/012071

[7] Dmour I, Taha MO. Novel nanoparticles based on chitosan-dicarboxylate conjugates via tandem ionotropic/covalent crosslinking with tripolyphosphate and subsequent evaluation as drug delivery vehicles. International Journal of Pharmaceutics. 2017;**529**(1–2):15-31. DOI: 10.1016/j.ijpharm.2017.06.061 Epub 2017 Jun 19

[8] Mahmood T, Ullah A, Ali R. Improved nanocomposite materials and their applications. In: Sharma A, editor. Nanocomposite Materials for Biomedical and Energy Storage Applications. London: IntechOpen; 2022. DOI: 10.5772/intechopen.102538

[9] Elnashar MMM. Review article: Immobilized molecules using biomaterials and nanobiotechnology. Journal of Biomaterials and Nanobiotechnology. 2010;**1**(1):61-77. DOI: 10.4236/jbnb.2010.11008

[10] Danial E, Elnashar M, Awad G. Immobilized inulinase on grafted alginate beads prepared by the one-step and the two-steps methods. Industrial and Engineering Chemistry Research. 2010;**49**(7):3120-3125. DOI: 10.1021/ie100011z

[11] Elnashar MM, Danial EN, Awad GEA. Novel carrier of grafted alginate for covalent immobilization of inulinase. Industrial & Engineering Chemistry Research. 2009;**48**(22): 9781-9785. DOI: 10.1021/ie9011276

[12] Elnashar MM, Millner PA, Johnson AF, Gibson TD. Parallel plate equipment for preparation of uniform gel sheets. Biotechnology Letters. 2005; **27**(10):737-739. DOI: 10.1007/s10529-005-5363-0

[13] Jacob E, Borah A, Jindal A, Pillai S, Yamamoto Y, Maekawa T, et al. Synthesis and characterization of citrus-derived pectin nanoparticles based on

their degree of esterification. Journal of Materials Research. 2020;**35**(12): 1514-1522. DOI: 10.1557/jmr.2020.108

[14] Feng Z, Xu J, Ni C. Preparation of redox responsive modified xanthan gum nanoparticles and the drug controlled release. International Journal of Polymeric Materials and Polymeric Biomaterials. 2021;**70**(14):994-1001. DOI: 10.1080/00914037.2020.1767618

[15] Gagliardi A, Giuliano E, Venkateswararao E, Fresta M, Bulotta S, Awasthi V, et al. Biodegradable polymeric nanoparticles for drug delivery to solid tumors. Frontiers in Pharmacology. 2021;**12**:601626. DOI: 10.3389/fphar.2021.601626

[16] Cazón P, Cazón D, Vázquez M, Guerra-Rodriguez E. Rapid authentication and composition determination of cellulose films by UV-VIS-NIR spectroscopy. Food Packaging and Shelf Life. 2022;**31**:1-9. DOI: 10.1016/j.fpsl.2021.100791

[17] Sultana S, Ahmad N, Faisal SM, Owais M, Sabir S. Synthesis, characterisation and potential applications of polyaniline/chitosan-Ag-nano-biocomposite. IET Nanobiotechnology. 2017;**11**(7):835-842. DOI: 10.1049/iet-nbt.2016.0215

[18] Niroomand F, Khosravani A, Younesi H. Fabrication and properties of cellulose-nanochitosan biocomposite film using ionic liquid. Cellulose. 2016; **23**:1311-1324

[19] Lara BRB, de Andrade PS, Guimarães Junior M, et al. Novel whey protein isolate/ polyvinyl biocomposite for packaging: Improvement of mechanical and water barrier properties by incorporation of nano-silica. Journal of Polymers and the Environment. 2021;**29**:2397-2408. DOI: 10.1007/s10924-020-02033-x

[20] Mahmoud AA, Osman O, Eid K, Ashkar E, Okasha A, Atta D, et al. FTIR spectroscopy of natural biopolymers blends. Middle East Journal of Applied Sciences. 2014;**4**(4):816-824. ISSN: 2077-4613

[21] Beasley M, Bartelink E, Taylor L, Miller R. Comparison of transmission FTIR, ATR, and DRIFT spectra: Implications for assessment of bone bioapatite diagenesis. Journal of Archaeological Science. 2014;**46**:16-22, ISSN 0305-4403. DOI: 10.1016/j.jas.2014.03.008

[22] Gieroba B, Sroka-Bartnicka A, Kazimierczak P, Kalisz G, Lewalska-Graczyk A, Vivcharenko V, et al. Surface chemical and morphological analysis of chitosan/1,3-β-d-Glucan polysaccharide films cross-linked at 90°C. International Journal of Molecular Sciences. 2022; **23**(11):5953. DOI: 10.3390/ijms23115953

[23] Van de Velde K, Kiekens P. Structure analysis and degree of substitution of chitin, chitosan and dibutyrylchitin by FT-IR spectroscopy and solid state 13C NMR. Carbohydrate Polymers. 2004; **58**(4):409-416, ISSN 0144-8617. DOI: 10.1016/j.carbpol.2004.08.004

[24] Qu Z, Schueneman G, Shofner ML, Meredith JC. Acrylic functionalization of cellulose nanocrystals with 2-isocyanatoethyl methacrylate and formation of composites with poly (methyl methacrylate). ACS Omega. 2020;**5**(48):31092-31099. DOI: 10.1021/acsomega.0c04246

[25] Huamani-Palomino RG, Córdova BM, Pichilingue LER, Venâncio T, Valderrama AC. Functionalization of an alginate-based material by oxidation and reductive amination. Polymers. 2021;**13**(2):255. DOI: 10.3390/polym13020255

[26] Cheaburu-Yilmaz C, Karavana S, Yilmaz O. Grafted copolymer based on chitosan and poly(n-isopropylacryl amide) via click technique. I. Synthesis and characterization. WSEAS Transactions on Biology and Biomedicine. 2017;**14**:120-128

[27] Gomez-Maldonado D, Filpponen I, Hernandez-Diaz JA, Waters MN, Auad ML, Johansson L-S, et al. Simple functionalization of cellulose beads with pre-propargylated chitosan for clickable scaffold substrates. Cellulose. 2021; **28**(10):6073-6087. DOI: 10.1007/s10570-021-03905-8

[28] Alirezvani ZG, Dekamin M, Davoodi F, Valiey E. Melamine-functionalized Chitosan: A new bio-based reusable bifunctional organocatalyst for the synthesis of cyanocinnamonitrile intermediates and densely functionalized nicotinonitrile derivatives. Chemistry Select. 2018;**3**: 10450. DOI: 10.1002/slct.201802010

[29] Stanescu PO, Radu IC, Leu Alexa R, Hudita A, Tanasa E, Ghitman J, et al. Novel chitosan and bacterial cellulose biocomposites tailored with polymeric nanoparticles for modern wound dressing development. Drug Delivery. 2021;**28**(1):1932-1950. DOI: 10.1080/10717544.2021.1977423

[30] İlgü H, Turan T, Şanli-Mohamed G. Preparation, characterization and optimization of chitosan nanoparticles as carrier for immobilization of thermophilic recombinant esterase. Journal of Macromolecular Science, Part A. 2011;**48**(9):713-721. DOI: 10.1080/10601325.2011.596050

[31] Cheng HN, English AD. Advances in the NMR spectroscopy of polymers: An overview. In: NMR Spectroscopy of Polymers in Solution and in the Solid State. Chapter 1. 2002. pp. 3-20. DOI: 10.1021/bk-2003-0834.ch001

[32] Porcino M, Li X, Gref R, Martineau-Corcos C. Solid-state NMR spectroscopy: A key tool to unravel the supramolecular structure of drug delivery systems. Molecules. 2021;**26**(14):4142. DOI: 10.3390/molecules26144142

[33] El Hariri El Nokab M, Habib MH, Alassmy YA, Abduljawad MM, Alshamrani KM, Sebakhy KO. Solid state NMR a powerful technique for investigating sustainable/renewable cellulose-based materials. Polymers 2022; **14**(5):1049. DOI: 10.3390/polym14051049

[34] Reich HJ. 8.1 Relaxation in NMR Spectroscopy. Wisconsin , USA: University of Wisconsin; 7th Aug. 2017. Available from:. p. www.chem.wisc.edu/areas/reich/nmr/08-tech-01-relax.htm

[35] Allert RD, Briegel KD, Bucher DB. Advances in nano- and microscale NMR spectroscopy using diamond quantum sensors. Chemical Communications. 2022;**58**(59):8165-8181. DOI: 10.1039/d2cc01546c

[36] Lin Y-J, Chuang W-T, Hsu S-H. Gelation mechanism and structural dynamics of chitosan self-healing hydrogels by in situ SAXS and coherent X-ray scattering. ACS Macro Letters. 2019;**8**:1449-1455. DOI: 10.1021/acsmacrolett.9b00683

[37] Heinze T, Haack V, Rensing S. Starch derivatives of high degree of functionalization. 7. Preparation of Cationic 2-hydroxypropyltrimethylammonium chloride starches. Starch/Stärke. 2004;**56**: 288-296. DOI: 10.1002/star.200300243

[38] Hamed AA, Abdelhamid IA, Saad GR, Elkady NA, Elsabee MZ. Synthesis, characterization and

antimicrobial activity of a novel chitosan Schiff bases based on heterocyclic moieties. International Journal of Biological Macromolecules. 2020;**153**: 492-501. DOI: 10.1016/j. ijbiomac.2020.02.302 Epub 2020 Feb 26

[39] Zhou N, Pan F, Ai X, Tuersuntuoheti T, Zhao L, Zhao L, et al. Preparation, characterization and antioxidant activity of sinapic acid grafted chitosan and its application with casein as a nanoscale delivery system for black rice anthocyanins. International Journal of Biological Macromolecules. 2022;**210**:33-43. DOI: 10.1016/j. ijbiomac.2022.05.010 Epub 2022 May 6

[40] Craciun AM, Morariu S, Marin L. Self-healing chitosan hydrogels: Preparation and rheological characterization. Polymers (Basel). 2022; **14**(13):2570. DOI: 10.3390/ polym14132570

[41] Wang Y, Karmakar T, Ghosh N, Basak S, Gopal SN. Targeting mangiferin loaded N-succinyl chitosan-alginate grafted nanoparticles against atherosclerosis - A case study against diabetes mediated hyperlipidemia in rat. Food Chemistry. 2022;**370**:131376. DOI: 10.1016/j.foodchem.2021.131376 Epub 2021 Oct 9

[42] Venkateshaiah A, Rajender N, Suresh, K. Chapter 14 - X-ray diffraction spectroscopy of polymer nanocomposites. In: Thomas S, Rouxel D, Ponnamma D, editors. Spectroscopy of Polymer Nanocomposites. New York, USA: William Andrew Publishing. 2016. p. 410-451. ISBN 9780323401838, DOI: 10.1016/ B978-0-323-40183-8.00014-8

[43] Burapapadh K, Takeuchi H, Sriamornsak P. Development of pectin nanoparticles through mechanical homogenization for dissolution enhancement of itraconazole. Asian Journal of Pharmaceutical Sciences. 2016;**11**(3):365-375, ISSN 1818-0876. DOI: 10.1016/j.ajps.2015.07.003

[44] Ahmad M, Gani A, Hassan I, Huang Q, Shabbir H. Production and characterization of starch nanoparticles by mild alkali hydrolysis and ultra-sonication process. Scientific Reports. 2020;**10**(1):3533. DOI: 10.1038/ s41598-020-60380-0

[45] Son D, Cho S, Nam J, Lee H, Kim M. X-ray-based spectroscopic techniques for characterization of polymer nanocomposite materials at a molecular level. Polymers (Basel). 2020;**12**(5):1053. DOI: 10.3390/polym12051053

[46] Shahid N, Erum A, Zaman M, Tulain UR, Shoaib QU, Malik NS, et al. Synthesis and evaluation of chitosan based controlled release nanoparticles for the delivery of ticagrelor. Designed Monomers and Polymers. 2022;**25**(1):55-63. DOI: 10.1080/15685551.2022.2054117

[47] Proctor S, Lovera S, Tomich A, Lavallo V. Searching for the truth: Elemental analysis-a powerful but often poorly executed technique. ACS Central Science. 2022;**8**(7):874-876. DOI: 10.1021/acscentsci.2c00761

[48] Li J, Ma FK, Dang QF, Liang XG, Chen XG. Glucose-conjugated chitosan nanoparticles for targeted drug delivery and their specific interaction with tumor cells. Frontiers of Materials Science. 2014;**8**:363-372. DOI: 10.1007/ s11706-014-0262-8

[49] Mishra RK, Sutar PB, Singhal JP, Banthia AK. Graft copolymerization of pectin with polyacrylamide. Polymer-Plastics Technology and Engineering. 2007;**46**(11):1079-1085. DOI: 10.1080/ 03602550701525164

[50] Zheng X-F, Lian Q, Yang H, Zhu H. Alkyl pectin: Hydrophobic matrices for

controlled drug release. Journal of Applied Polymer Science. 2015;**132**: 41302. DOI: 10.1002/app.41302

[51] Zechuan Y, Fan L, Qingrong H, Guo Z, Tongfei S. Synthesis and properties of the amino acid functionalized curcumin/his-pectin colloidal particles. Chemical Journal of Chinese Universities. 2016;**37**(2):381. DOI: 10.7503/cjcu20150591

[52] Espino-Pérez E, Domenek S, Belgacem N, Sillard C, Bras J. Green process for chemical functionalization of nanocellulose with carboxylic acids. Biomacromolecules. 2014;**15**(12): 4551-4560. DOI: 10.1021/bm5013458 Epub 2014 Nov 12

[53] Boujemaoui A, Mongkhontreerat S, Malmström E, Carlmark A. Preparation and characterization of functionalized cellulose nanocrystals. Carbohydrate Polymers. 2015;**115**:457-464. DOI: 10.1016/j.carbpol.2014.08.110

[54] Nasrabadi M, Beyramabadi SA, Morsali A. Surface functionalization of chitosan with 5-nitroisatin. International Journal of Biological Macromolecules. 2020;**147**:534-546. DOI: 10.1016/j. ijbiomac.2020.01.070 Epub 2020 Jan 11

[55] Taubner T, Marounek M, Synytsya A. Preparation and characterization of amidated derivatives of alginic acid. International Journal of Biological Macromolecules. 2017;**103**:202-207. DOI: 10.1016/j.ijbiomac.2017.05.070 Epub 2017 May 17

[56] Chang R, Tian Y, Yu Z, Sun C, Jin Z. Preparation and characterization of zwitterionic functionalized starch nanoparticles. International Journal of Biological Macromolecules. 2020; **1**(142):395-403. DOI: 10.1016/j. ijbiomac.2019.09.110 Epub 2019 Oct 14

[57] Cazotti JC, Fritz AT, Garcia-Valdez O, Smeets NMB, Dubé MA, Cunningham MF. Graft modification of starch nanoparticles using nitroxide-mediated polymerization and the grafting to approach. Biomacromolecules. 2020; **21**(11):4492-4501. DOI: 10.1021/acs. biomac.0c00462 Epub 2020 May 14

[58] Corcione CE, Frigione M. Characterization of nanocomposites by thermal analysis. Materials (Basel). 2012; **5**(12):2960-2980. DOI: 10.3390/ ma5122960

[59] Kumar A, Singh P, Nanda A. Hot stage microscopy and its applications in pharmaceutical characterization. Applied Microscopy. 2020;**50**(1):12. DOI: 10.1186/s42649-020-00032-9

[60] Neto JSS, de Queiroz HFM, Aguiar RAA, Banea MD. A review on the thermal characterisation of natural and hybrid fiber composites. Polymers (Basel). 2021;**13**(24):4425. DOI: 10.3390/ polym13244425

[61] Pielichowska K, Nowicka K. Analysis of nanomaterials and nanocomposites by thermoanalytical methods. Thermochimica Acta. 2019;**675**:140-163, ISSN 0040-6031. DOI: 10.1016/j. tca.2019.03.014

[62] Xiao P, Zhang J, Ye F, Wu J, He J, Zhang J. Synthesis, characterization and properties of novel cellulose derivatives containing phosphorus: Cellulose diphenyl phosphate and its mixed esters. Cellulose. 2014;**21**:2369-2378. DOI: 10.1007/s10570-014-0256-9

[63] Hu Y, Hu S, Zhang S, Dong S, Hu J, Kang L, et al. A double-layer hydrogel based on alginate-carboxymethyl cellulose and synthetic polymer as sustained drug delivery system. Scientific Reports. 2021;**11**(1):9142. DOI: 10.1038/s41598-021-88503-1

[64] Kahdestani SA, Shahriari MH, Abdouss M. Synthesis and characterization of chitosan nanoparticles containing teicoplanin using sol–gel. Polymer Bulletin. 2021;**78**: 1133-1148. DOI: 10.1007/s00289-020-03134-2

[65] Kassab Z, Aziz F, Hannache H, Ben Youcef H, El Achaby M. Improved mechanical properties of k-carrageenan-based nanocomposite films reinforced with cellulose nanocrystals. International Journal of Biological Macromolecules. 2019;**123**:1248-1256. DOI: 10.1016/j.ijbiomac.2018.12.030 Epub 2018 Dec 4

[66] Abasalizadeh F, Moghaddam SV, Alizadeh E, et al. Alginate-based hydrogels as drug delivery vehicles in cancer treatment and their applications in wound dressing and 3D bioprinting. Journal of Biological Engineering. 2020; **14**:8. DOI: 10.1186/s13036-020-0227-7

[67] Maiti S, Jana S. Biocomposites in ocular drug delivery. In: Jana S, Maiti S, editors. Biopolymer-Based Composites. Cambridge, UK: Woodhead Publishing. 2017. pp. 139-168. ISBN 9780081019146. DOI: 10.1016/B978-0-08-101914-6.00006-5

[68] da Silva ISV, Neto WPF, Silvério HA, Pasquini D, Zeni Andrade M, Otaguro H. Mechanical, thermal and barrier properties of pectin/cellulose nanocrystal nanocomposite films and their effect on the storability of strawberries (Fragaria ananassa). Polymers for Advanced Technologies. 2017;**28**:1005-1012. DOI: 10.1002/pat.3734

[69] Crut A, Maioli P, Del Fatti N, Vall'ee, F. Optical absorption and scattering spectroscopies of single nano-objects. Chemical Society Reviews. 2014;**43**:3921

[70] Wan Q. Scanning electron microscopy investigation of bio-polymer composites morphology. [PhD thesis], University of Sheffield. 2017

[71] Kaushik M, Fraschini C, Chauve G, Moores JPA. Transmission electron microscopy for the characterization of cellulose nanocrystals. In: Maaz K. editor. The Transmission Electron Microscope - Theory and Applications. London: IntechOpen; 2015. DOI: 10.5772/60985

[72] Venkateshaiah A, Padil VVT, Nagalakshmaiah M, Waclawek S, Černík M, Varma RS. Microscopic techniques for the analysis of micro and nanostructures of biopolymers and their derivatives. Polymers (Basel). 2020; **12**(3):512. DOI: 10.3390/polym12030512

[73] Malik NS, Ahmad M, Minhas MU, Tulain R, Barkat K, Khalid I, et al. Chitosan/Xanthan gum based hydrogels as potential carrier for an antiviral drug: Fabrication, characterization, and safety evaluation. Frontiers in Chemistry. 2020;**8**:50. DOI: 10.3389/fchem.2020.00050

[74] Ansari F, Salajková M, Zhou Q, Berglund LA. Strong surface treatment effects on reinforcement efficiency in biocomposites based on cellulose nanocrystals in poly(vinyl acetate) matrix. Biomacromolecules. 2015;**16**(12): 3916-3924. DOI: 10.1021/acs.biomac.5b01245 Epub 2015 Nov 17

[75] Ulaganathan RK, Senusi NM, Mohamed Noor A, Wan Abdullah WN, Mohd Amin MA, Abdul Razab MK, et al. Effect of Cellulose Nanofibers (CNF) as Reinforcement in Polyvinyl Alcohol/CNF Biocomposite. Journal of Physics Conference Series. 2021;**2129**:012057

[76] Abral H, Kadriadi, Mahardika M, Handayani D, Sugiarti E, Muslimin AN. Characterization of disintegrated bacterial cellulose nanofibers/PVA

bionanocomposites prepared via ultrasonication. International Journal of Biological Macromolecules. 2019;**135**: 591-599. DOI: 10.1016/j. ijbiomac.2019.05.178 Epub 2019 May 25

[77] Oviedo M, Montoya Y, Agudelo W, García-García A, Bustamante J. Effect of molecular weight and nanoarchitecture of chitosan and polycaprolactone electrospun membranes on physicochemical and hemocompatible properties for possible wound dressing. Polymers. 2021;**13**:4320. DOI: 10.3390/ polym13244320

[78] Chiaoprakobkij N, Suwanmajo T, Sanchavanakit N, Phisalaphong M. Curcumin-loaded bacterial cellulose/ alginate/gelatin as a multifunctional biopolymer composite film. Molecules. 2020;**25**(17):3800. DOI: 10.3390/ molecules25173800

[79] Dmour I, Muti H. Application of dual ionic/covalent crosslinking in lecithin/chitosan nanoparticles and their evaluation as drug delivery system. Acta Poloniae Pharmaceutica. 2021;**78**:83-96. DOI: 10.3390/gels8080494

[80] Tucker IM, Corbett W, Fatkin JC, Jack J, Kaszuba RO, MacCreath M, et al. Laser Doppler electrophoresis applied to colloids and surfaces. Current Opinion in Colloid & Interface Science. 2015;**20**(4): 215-226. ISSN 1359-0294. DOI: 10.1016/j. cocis.2015.07.001

[81] Ramos AP. 4 - Dynamic Light Scattering Applied to Nanoparticle Characterization. In: Da Róz AL, Ferreira M, de Lima Leite F, Oliveira ON, editors. Nanocharacterization Techniques. Micro and Nano Technologies. William Andrew Publishing; 2017. pp. 99-110. DOI: 10.1016/ B978-0-323-49778-7.00004-7

[82] Rasmussen MK, Pedersen JN, Marie R. Size and surface charge characterization of nanoparticles with a salt gradient. Nature Communications. 2020;**11**:2337. DOI: 10.1038/s41467-020-15889-3

[83] Skoglund S, Hedberg J, Yunda E, Godymchuk A, Blomberg E, Odnevall WI. Difficulties and flaws in performing accurate determinations of zeta potentials of metal nanoparticles in complex solutions-Four case studies. PLoS One. 2017;**12**(7):e0181735. DOI: 10.1371/journal.pone.0181735

[84] Perera YR, Hill RA, Fitzkee NC. Protein interactions with nanoparticle surfaces: Highlighting solution NMR techniques. Israel Journal of Chemistry. 2019;**59**(11–12):962-979. DOI: 10.1002/ ijch.201900080

[85] Riegger BR, Bäurer B, Mirzayeva A, Tovar GEM, Bach M. A systematic approach of chitosan nanoparticle preparation via emulsion crosslinking as potential adsorbent in wastewater treatment. Carbohydrate Polymers. 2018;**180**:46-54. DOI: 10.1016/j. carbpol.2017.10.002

[86] Gu H, Gao X, Zhang H, Chen K, Peng L. Fabrication and characterization of cellulose nanoparticles from maize stalk pith via ultrasonic-mediated cationic etherification. Ultrasonics Sonochemistry. 2020;**66**:1-10, Article 104932. DOI: 10.1016/j. ultsonch.2019.104932

[87] Dmour I, Islam N. Recent advances on chitosan as an adjuvant for vaccine delivery. International Journal of Biological Macromolecules. 2022;**200**: 498-519. DOI: 10.1016/j. ijbiomac.2021.12.129

[88] Gunathilake TMSU, Ching YC, Uyama H, et al. Enhanced curcumin loaded nanocellulose: a possible inhalable nanotherapeutic to treat COVID-19.

Cellulose. 2022;**29**:1821-1840. DOI: 10.1007/s10570-021-04391-8

[89] Bezerra RDS, Silva MMF, Morais AIS, Osajima JA, Santos MRMC, Airoldi C, et al. Phosphated cellulose as an efficient biomaterial for aqueous drug ranitidine removal. Materials (Basel). 2014;7(12):7907-7924. DOI: 10.3390/ma7127907

[90] Fan H, Ma Y, Wan J, et al. Adsorption properties and mechanisms of novel biomaterials from banyan aerial roots via simple modification for ciprofloxacin removal. The Science of the Total Environment. 2020;**708**: 134630. DOI: 10.1016/j.scitotenv.2019.134630

[91] Oshima T, Taguchi S, Ohe K, Baba Y. Phosphorylated bacterial cellulose for adsorption of proteins. Carbohydrate Polymers. 2011;**83**:953-958. DOI: 10.1016/j.carbpol.2010.09.005

[92] Lee DW, Shirley SA, Lockey RF, Mohapatra SS. Thiolated chitosan nanoparticles enhance anti-inflammatory effects of intranasally delivered theophylline. Respiratory Research. 2006;7(1):112. DOI: 10.1186/1465-9921-7-112

[93] Das S, Ghosh B, Sarkar K. Nanocellulose as sustainable biomaterials for drug delivery. Sensors International. 2022;**3**:100135, ISSN 2666-3511. DOI: 10.1016/j.sintl.2021.100135

[94] Ramachandran S, Narasimman V, Rajesh P. Low molecular weight sulfated chitosan isolation, characterization and anti-tuberculosis activity derived from *Sepioteuthis lessoniana*. International Journal of Biological Macromolecules. 2022;**206**:29-39. DOI: 10.1016/j.ijbiomac.2022.02.121 Epub 2022 Feb 24

[95] Stojkov G, Niyazov Z, Picchioni F, Bose R. Relationship between structure and rheology of hydrogels for various applications. Gels. 2021;7(4):255. DOI: 10.3390/gels7040255

[96] Nilsen-Nygaard J, Strand SP, Vårum KM, Draget KI, Nordgård CT. Chitosan: Gels and interfacial properties. Polymers. 2015;7(3):552-579. DOI: 10.3390/polym7030552

[97] Zhao J, Wang Y, Luo G, Zhu S. Immobilization of penicillin G acylase on macro-mesoporous silica spheres. Bioresource Technology. 2011;**102**(2): 529-535. DOI: 10.1016/j.biortech.2010.09.076 Epub 2010 Sep 27

[98] Elnashar MM, Yassin MA, Abdel Moneim AE-F, Abdel Bary EM. Surprising performance of alginate beads for the release of low-molecular-weight drugs. Journal of Applied Polymer Science. 2010;**116**:3021-3026. DOI: 10.1002/app.31836

[99] Lingayya H, Shaila V, Shivaprada HS, Anitha GS. Development and characterization of pectin and chitosan based biocomposite material for bio-medical application. International Journal of Material Sciences & Engineering. 2021;**11**:109 2169-0022

[100] Ahsan A, Farooq MA, Parveen A. Thermosensitive chitosan-based injectable hydrogel as an efficient anticancer drug carrier. ACS Omega. 2020;**5**(32):20450-20460. DOI: 10.1021/acsomega.0c02548

[101] Liu T, Bolle ECL, Chirila TV, Buck M, Jonas D, Suzuki S, et al. Transparent, pliable, antimicrobial hydrogels for ocular wound dressings. Applied Sciences. 2020;**10**(21):7548. DOI: 10.3390/app10217548

[102] Demeter M, Călina I, Scărișoreanu A, Micutz M. E-beam cross-linking of complex hydrogels

formulation: The influence of poly (ethylene oxide) concentration on the hydrogel properties. Gels. 2021;**8**(1):27. DOI: 10.3390/gels8010027

[103] Chiaoprakobkij N, Seetabhawang S, Sanchavanakit N, Phisalaphong M. Fabrication and characterization of novel bacterial cellulose/alginate/gelatin biocomposite film. Journal of Biomaterials Science Polymer Edition. 2019;**30**:961-982. DOI: 10.1080/09205063.2019.1613292

[104] Ibrahim MM, Nair AB, Shehata BEATM. Hydrogels and their combination with liposomes, niosomes, or transfersomes for dermal and transdermal drug delivery. In: Catala A, editor. Liposomes. London: IntechOpen; 2017. DOI: 10.5772/intechopen.68158

[105] Du CC, Huang W. Progress and prospects of nanocomposite hydrogels in bone tissue engineering. Nano. 2022; **8**(1):102-124. DOI: 10.1080/20550324.2022.2076025

[106] Chaichi M, Badii F, Mohammadi A, Hashemi M. Water resistance and mechanical properties of low methoxy-pectin nanocomposite film responses to interactions of Ca2+ ions and glycerol concentrations as crosslinking agents. Food Chemistry. 2019;**293**:429-437. DOI: 10.1016/j.foodchem.2019.04.110

[107] Adam F, Jamaludin J, Abu Bakar SH, Abdul Rasid R, Hassan Z. Evaluation of hard capsule application from seaweed: Gum Arabic-Kappa carrageenan biocomposite films. Cogent Engineering. 2020;**7**(1):1765682. DOI: 10.1080/23311916.2020.1765682

[108] Kurowiak J, Mackiewicz A, Klekiel T, Będziński R. Evaluation of selected properties of sodium alginate-based hydrogel material—Mechanical strength, μDIC analysis and degradation. Materials. 2022;**15**:1225. DOI: 10.3390/ma15031225

[109] Gilbert L, Picard C, Savary G, Grisel M. Rheological and textural characterization of cosmetic emulsions containing natural and synthetic polymers: Relationships between both data. Colloids and Surfaces A: Physicochemical and Engineering Aspects. 2013;**421**:150-163. DOI: 10.1016/j.colsurfa.2013.01.003 hal-02507767

[110] Ogah OA. Rheological properties of natural fiber polymer composites. MOJ Polymer Science. 2017;**1**(4):147-148. DOI: 10.15406/mojps.2017.01.00022

[111] Dörr D, Kuhn U, Altstädt V. Rheological study of gelation and crosslinking in chemical modified polyamide 12 using a multiwave technique. Polymers. 2020;**12**(4):855. DOI: 10.3390/polym12040855

[112] Jannatamani H, Motamedzadegan A, Farsi M, Yousefi H. Rheological properties of wood/bacterial cellulose and chitin nano-hydrogels as a function of concentration and their nano-films properties. IET Nanobiotechnology. 2022; **16**(4):158-169. DOI: 10.1049/nbt2.12083 Epub 2022 Apr 4

[113] Ajovalasit A, Sabatino MA, Todaro S, Alessi S, Giacomazza D, Picone P, et al. Xyloglucan-based hydrogel films for wound dressing: Structure-property relationships. Carbohydrate Polymers. 2018;**179**: 262-272. DOI: 10.1016/j.carbpol.2017.09.092 Epub 2017 Sep 28

Chapter 3

Reactive Extrusion as an Environmentally Friendly Technology for the Production of Bio(Nano)Composites: Implementation and Characterization

Silvester Bolka and Blaž Nardin

Abstract

The influences of reactive extrusion of poly(lactic acid) (PLA)-based bio(nano) composites on their properties are described. Reactive compatibilizers were used to enable good dispersion of natural (nano)fibers in the thermoplastic matrix consisting of PLA/poly(butylene adipate-co-terephthalate) (PBAT) and PLA/polycarbonate (PC) blends. At the same time, chain extenders were used for the modification of immiscible thermoplastics, PLA and PBAT, in order to achieve good miscibility of the PLA/PBAT blend. In the experimental part, the main obstacle of PLA, its brittleness, was improved in three different series of bio(nano)composites. Reactive extrusion with PLA/PBAT blends and the addition of hops as a chain extender and compatibilizer increased the elongation at break of the bio(nano)composite by more than 240% and the impact strength by 200% compared to neat PLA. Reactive extrusion of PLA/PBAT blends and addition of 1% nanocrystalline cellulose (NCC) with additives increased the elongation at break by more than 730% compared to pure PLA, and the sample did not break during the impact testing. Reactive extrusion with PLA/PC blends and the addition of 1 wt% NCC with additives increased the elongation at break by more than 90% and the impact strength by more than 160% compared to pure PLA.

Keywords: bio(nano)composites, reactive compounding, hops fibers, NCC, characterization

1. Introduction

Biopolymers, biopolymer blends, and biocomposites are becoming more and more interesting for research and industry because they have less impact on the environment. Researchers are making great efforts to avoid the disadvantages of biopolymers. Environmentally friendly materials, especially biodegradable ones, such as PLA, poly(3-hydroxybutyrate) (PHB), poly(ε-caprolactone) (PCL), poly(butylene succinate) (PBA), and PBAT, are attracting great interest from researchers and industry [1]. The properties of PLA together with its processability on conventional equipment make it possible to replace conventional petroleum-based thermoplastics [2, 3]. There are numerous research efforts in the field of reactive modification of biopolymers using various reactive agents, such as organic peroxides and multi-functional coagents, for modification with crosslinking [4–6]. Researchers reported the degradation of PLA in combination with crosslinking by modifying PLA with peroxides [7–12]. Compatibilization can be used for the modification of biopolymers. Compatibilization of immiscible polymers can be performed by adding nonreactive agents, reactive agents, crosslinking, and double-functionalized polymers or with mechanochemistry where temperature and shear create macroradicals during compounding. Reactive compatibilization is a very cost-effective processing technology, an environmentally friendly process because it is solvent free, requires no special equipment, and can be easily up-scaled to industrial production [13]. A chain extender was used for the PLA/PBAT blend, which improved elongation at break, tensile strength, and impact resistance [14]. For the PLA/poly(3-hydroxybutyrate-co-3-hydroxyvalerate)/PBAT ternary blend, an epoxy-based styrene-acrylic oligomer with low functionality was used, which improved tensile strength and elongation at break [15]. For the PLA/PBAT blend, a bio-based chain extender (epoxidized cardanol prepolymer) was used, which improved tensile strength, elongation at break, and toughness [16]. Epoxy-functionalized oligomer as a chain extender was used for PLA/PBAT/flax fiber composites, where stiffness and strength were improved [17]. For microcellulose and nanocellulose, the cellulose surface is chemically modified to improve the surface interaction of cellulose with the polymer matrix, usually by esterification and silanization or by plasma and corona surface treatment, which is also required in the case of PLA matrix [18]. Unmodified bacterial cellulose nanowhiskers were incorporated into the PLA matrix by electrospinning, followed by the incorporation of nanostructured fiber into the PLA matrix by melt blending. The stiffness and strength were increased, while the ductility remained at the level of pure PLA [19]. Major research efforts have been devoted to improve the reactive compatibility of PLA, including with petroleum-based polymers. Chain extenders were used for the PLA/PC blend to increase toughness [20]. The tougher PLA base blend was mixed with PC, hydrogenated styrene-butadiene-styrene block copolymer and using a reactive compatibilizer and poly(ethylene-co-glycidyl methacrylate). Thermal stability and excellent toughness were achieved [21].

2. Materials and methods

2.1 Samples

Commercially available PLA with the trade name Ingeo 4043D was provided by Plastika Trček, Slovenia. A commercially available PC with the trade name Lexan 243 R was purchased from the company Sabic, Austria. NCC was donated by the company Navitas, Slovenia. Commercially available SEBS-*g*-MA with the trade name FG 1901 GT was purchased from Kraton, Germany. Commercially available TPU copolymer with trade name Kuramiron U TU-S5265 was purchased from Kuraray, Germany. Commercially available $CaCO_3$ with the trade name Calplex Extra was donated by

Sample	PLA (wt%)	PBAT (wt%)	TPU (wt%)	Joncryl (wt%)	NCC (wt%)
PLA	100	0	0	0	0
PLA15PBAT10H	70	15	4.5	0.5	10
PLA15PBAT5H	75	15	4.5	0.5	5
PLA20PBAT5H	70	20	4.5	0.5	5

Table 1.
Composition of the samples of the first series.

Calcit, Slovenia. A commercially available chain extender with the trade name Joncryl ADR 4468 was purchased from the company BASF, Netherlands. Commercially available PBAT with the trade name Ecoflex F Blend C1200 was purchased from the company BASF, Netherlands. Commercially available hops with the trade name Styrian Aurora were donated by the Slovenian Institute of Hop Research and Brewing, Slovenia.

Three different series of samples were produced. The first series was a toughness modification of PLA by blending with PBAT, adding a Joncryl chain extender to improve the miscibility of PLA and PBAT, and adding a modified TPU compatibilizer to improve the interactions between the hops and the thermoplastic matrix, since hops without surface treatment were used. The composition of the samples of the first series is shown in **Table 1**.

For the second series of samples, the toughest version from the first batch of samples was used, to which 1 wt% NCC was added. The composition of the second series of samples is shown in **Table 2**.

For the third series of samples, PLA was blended with PC to increase toughness and maintain stiffness and strength at a high level. NCC was added to the thermoplastic blend, to which two compatibilizers were added, at three different concentrations. In addition to the modified TPU, modified SEBS was also used to maximize the toughness of the bio(nano)composite because it has a high content of PC and the toughness is limited. The reactive compounding was performed twice. The composition of the third series of samples is shown in **Table 3**.

2.2 Reactive compounding

Reactive compounding was used to improve the surface interaction of NCC and hops with the thermoplastic matrix. The NCC and hops used were not surface-treated. The role of the compatibilizer was to ensure good surface interaction of the NCC and hops with the thermoplastic matrix and to ensure good dispersion of the

Sample	PLA (wt%)	PBAT (wt%)	TPU (wt%)	Joncryl (wt%)	NCC (wt%)
PLA20PBAT	74.5	20	5	0.5	0
PLA20PBAT 1NCC-1	73.5	20	5	0.5	1

Table 2.
Composition of the samples of the second series.

Sample	PLA (wt%)	PC (wt%)	SEBS (wt%)	TPU (wt%)	$CaCO_3$ (wt%)	NCC (wt%)	Compounding cycles
PLAPC	42	40	10	5	3	0	1
PLAPC 1NCC-1	41	40	10	5	3	1	1
PLAPC 1NCC-2	41	40	10	5	3	1	2
PLAPC 2NCC-1	40	40	10	5	3	2	1
PLAPC 2NCC-2	40	40	10	5	3	2	2
PLAPC 5NCC-1	37	40	10	5	3	5	1
PLAPC 5NCC-2	37	40	10	5	3	5	2

Table 3.
Composition of the samples of the third series and the number of compounding cycles.

NCC and hops in the thermoplastic matrix by qualitatively wetting the surface of the NCC and hops to prevent its agglomeration. To ensure good wettability of the NCC surface and its dispersion, high shear was used in reactive compounding and, in the case of the PLAPC samples, multiple compounding cycles were used. High shear was achieved by high screw speeds during reactive compounding and the lowest possible processing temperature for the bio(nano)composites. In parallel, multiple reactive compounding cycles can be used to mimic the recycling of bio(nano)composites. The behavioral changes during multiple processing of bio(nano)composites can be studied. Reactive compounding is an existing technology for modifying PLA. Thus, the main drawback of PLA, namely its brittleness, can be improved by reactive compounding by preparing a blend of PLA and a tough thermoplastic with the addition of natural fibers and a reactive additive.

For the first reactive extrusion cycle, the materials were mixed separately and extruded on the Labtech LTE 20–44 twin-screw extruder. The screw diameter was 20 mm, the L/D ratio was 44:1, and the screw speed was 600 rpm. The temperature profile for the PLAPC and PLAPBAT samples increased from the hopper (165°C and 145°C, respectively) to the die (200°C and 180°C, respectively). Vacuum extraction was performed during reactive extrusion to remove the volatile gaseous products of reactive extrusion. The vacuum was set at 50 mbar. After compounding, the two produced filaments with a diameter of 3 mm were cooled in a water bath and formed into pellets with a length of about 5 mm and a diameter of 3 mm.

In the case of the second reactive extrusion cycle in the samples PLAPC, the produced pellets of bio(nano)composites were extruded on the same extruder with identical extruder settings.

2.3 Injection molding

Injection molding was performed on Krauss Maffei 50–180 CX injection molding machine with a screw diameter of 30 mm and a clamping force of 500 kN. The cold runner mold was used to produce the samples. The mold had two cavities, one with a dumbbell-shaped mold of type 1BA (ISO 527-1), and the second with a cuboid shape (ISO 178/ISO 179). The temperature profile for the PLAPC and PLAPBAT samples was increasing from the hopper (185°C and 165°C, respectively) to the mold (200°C and 185°C, respectively), the injection speed was set to 60 mm/min, and the mold temperature was set to 30°C and 20°C, respectively, and the cooling time was set to 10 s and 15 s, respectively. During plastification, the backpressure for the PLAPC and PLAPBAT samples was set to 150 bar and 250 bar, respectively, and the screw speed was set to 50 rpm and 200 rpm, respectively. The high backpressure was used to remove air pockets in the melt and to achieve the best possible homogeneity of the melt. For the PLAPC samples, a low screw speed was used to prevent thermal degradation of the bio(nano)composite melt and to minimize shear during processing due to the higher processing temperature of the PLAPC samples.

2.4 Methods for characterization of the bio(nano)composites

Flexural and tensile tests were performed on the Shimadzu AG-X plus according to ISO 178 and ISO 527-1, respectively. Five measurements were taken for each specimen. In tensile tests, tensile stiffness (E_t), tensile strength (σ_m), tensile yield strain (ε_m), and elongation at break (ε_{tb}) were determined. In bending, the flexural stiffness (E_f), flexural strength (σ_{fM}), and yield strain (ε_{fM}) were evaluated.

Thermomechanical properties were investigated using a Perkin Elmer DMA 8000 dynamic mechanical analyzer. TT_DMA software, version 14,310, was used to evaluate the results. The viscoelastic properties of the samples were analyzed by recording the storage modulus (E'), loss modulus (E"), and loss factor (tan δ) as a function of temperature. The viscoelastic analyses were performed on specimens with dimensions of approximately 42 x 5 x 2 mm. The samples were heated at 2°C/min from room temperature (23°C) to 180°C under an air atmosphere. A frequency of 1 Hz and an amplitude of 20 μm were used in dual-cantilever mode.

Thermal measurements were performed using a differential scanning calorimeter (DSC 2, Mettler Toledo) under a nitrogen atmosphere (20 mL/min). The temperature of the samples was raised from 0 to 200°C at a heating rate of 10°C/min and held in the molten state for 5 min to erase their thermal history. After cooling at 10°C/min, the samples were reheated at 200°C at 10°C/min. The crystallization temperature (T_c), crystallization enthalpy (ΔH_c), glass transition temperature (T_g), cold crystallization temperature (T_{cc}), cold crystallization enthalpy (ΔH_{cc}), melting temperature (T_m), and melting enthalpy (ΔH_m) were determined using the cooling and the second heating scan.

Crystallization behavior on samples PLAPC was determined on Mettler Toledo Flash DSC 1 with Huber intercooler TC45 and nitrogen purge gas (50 mL/min). Samples were cooled from melt (200°C) to the desired temperature, rapidly heated to the aging temperature (90°C for 100 s), rapidly cooled to 15°C, reheated at 120°C, and then cold crystallized at 120°C at various times (from 0.1 s to 2400 s). All cooling and heating segments were rapidly cooled and heated (500°C/s) to prevent crystallization during cooling and heating. The first heating run was performed from 15–200°C. For the evaluation of the heating section, segment No. 12 was taken and the melting temperature and melting enthalpy were characterized. The mass of the samples was determined using the normalized change in specific heat capacity at the glass transition based on the evaluation of DSC 2 measurements.

Impact tests were performed on Pendel Dongguan Liyi test equipment, type LY -XJJD5 impact testing machine according to ISO 179. The impact test specimens were injection molded according to ISO 179 and had dimensions of 80 x 10 x 4 mm. The pendulum with 5 J was used for the evaluation of the impact test.

3. Results and discussion

3.1 Mechanical properties

The tensile and flexural results are shown in **Table 4** and **Figure 1**. The results for the PLA sample are used as reference values for all other modifications by reactive compounding and for evaluating the usability of reactive compounding.

3.1.1 Results of the first-reactive compounding series

When hops were added to the blend of PLA and PBAT, increasing hops decreased tensile stiffness, strength and elongation at break, slightly increased flexural stiffness, decreased flexural strength, and flexural elongation. Simultaneously decreasing the hops content and increasing the PBAT content had no effect on tensile stiffness and decreased strength, but dramatically increased elongation at break, decreased flexural stiffness and strength, and increased elongation at flexural strength. It can be concluded that the addition of hops to the biocomposites lowered the strength, tensile stiffness, flexural stiffness, and elongation. The addition of PBAT lowered the stiffness and strength, but dramatically increased the elongation at break. Adding PBAT to PLA can improve PLA's biggest drawback, its brittleness. The second conclusion is that reactive compounding for the combination of the thermoplastic matrix of PLA and PBAT modified with the chain extender and the compatibilizer with the addition

Sample	Tensile test results			Flexural test results		
	E_t (GPa)	σ_m (MPa)	ε_{tb} (%)	E_f (GPa)	σ_{fM} (MPa)	ε_{fM} (%)
PLA	3.17 ± 0.21	71.8 ± 1.0	4.8 ± 0.3	3.38 ± 0.07	105.2 ± 0.8	4.49 ± 0.04
PLA15PBAT 10H	2.47 ± 0.23	46.9 ± 0.7	3.9 ± 0.2	2.87 ± 0.02	71.5 ± 0.2	3.38 ± 0.01
PLA15PBAT 5H	2.59 ± 0.17	50.4 ± 0.5	8.9 ± 1.2	2.84 ± 0.01	74.1 ± 0.3	3.50 ± 0.01
PLA20PBAT 5H	2.45 ± 0.33	42.9 ± 0.8	16.5 ± 2.0	2.48 ± 0.01	64.3 ± 0.2	3.85 ± 0.04
PLA20PBAT	2.65 ± 0.24	44.7 ± 0.5	61.2 ± 10.6	2.16 ± 0.02	55.2 ± 0.4	3.74 ± 0.05
PLA20PBAT 1NCC	2.15 ± 0.15	44.1 ± 0.4	39.9 ± 5.3	2.13 ± 0.01	53.2 ± 0.3	3.84 ± 0.03
PLAPC	2.17 ± 0.16	31.6 ± 0.3	4.8 ± 0.6	2.03 ± 0.02	48.2 ± 1.1	3.13 ± 0.28
PLAPC 1NCC-1	2.37 ± 0.31	40.5 ± 0.2	9.6 ± 0.4	2.08 ± 0.01	57.6 ± 0.3	4.68 ± 0.08
PLAPC 1NCC-2	2.44 ± 0.24	40.7 ± 0.4	9.5 ± 0.8	2.02 ± 0.01	55.6 ± 0.3	4.71 ± 0.06
PLAPC 2NCC-1	2.38 ± 0.27	37.4 ± 0.2	8.7 ± 0.9	1.95 ± 0.01	52.6 ± 0.3	4.84 ± 0.11
PLAPC 2NCC-2	2.33 ± 0.11	37.3 ± 0.4	8.8 ± 1.0	1.88 ± 0.01	51.1 ± 0.2	4.86 ± 0.14
PLAPC 5NCC-1	2.23 ± 0.07	36.6 ± 0.5	8.6 ± 0.6	1.94 ± 0.01	51.6 ± 0.5	4.77 ± 0.13
PLAPC 5NCC-2	2.01 ± 0.13	31.5 ± 0.6	4.6 ± 0.3	1.81 ± 0.01	40.4 ± 2.4	3.00 ± 0.49

Table 4.
Summarized results from the tensile and flexural tests.

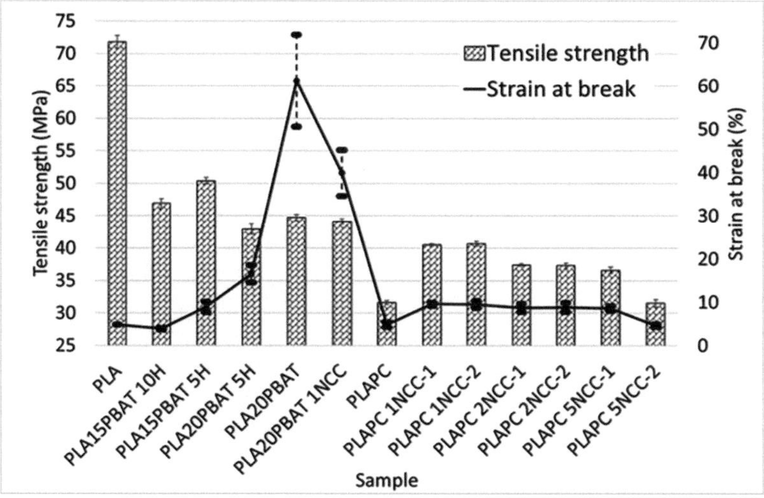

Figure 1.
Summarized results of the tensile strength (bars) and strain at break (line).

of hops is the right technological approach for the production of biocomposites. The miscibility of the thermoplastics was achieved by the correct choice of the chain extender at the appropriate concentration, as the elongation at break at 20 wt% addition of PBAT to PLA and simultaneous addition of 5 wt% hops increased dramatically compared to the other biocomposites. Compared to the reference values for pure PLA, the stiffness and strength decreased significantly, but at the same time the elongation at break and an indicator of toughness was drastically increased.

3.1.2 Results of the second-reactive compounding series

The addition of NCC was compared with the mixture PLA/PBAT as a reference for the second series of samples. The addition of 1 wt% NCC decreased the stiffness, strength, and elongation at break, but increased the flexural strain. Compared with the sample from the first series PLA20PBAT 5H, the stiffness of the sample PLA20PBAT 1NCC was lower, the strength was higher, and the elongation at break was much higher. The flexural properties were all lower. It can be concluded that the addition of 1 wt% NCC decreased the strength, stiffness, and elongation due to the poorer wettability of NCC. The processing conditions of bio(nano)composites are not optimal for the incorporation of NCC into the thermoplastic matrix. Nevertheless, the elongation at break of bio(nano)composites with NCC is significantly higher compared to the hops composite, indicating that NCC is a suitable additive to increase the toughness of PLA-based bio(nano)composites. Despite the nonoptimal reaction conditions, a drastically higher toughness was achieved compared to PLA/PBAT/hops biocomposites and also to the PLA reference.

3.1.3 Results of the third-reactive compounding series

The third series of samples was used to test NCC and the effect of multiple compounding cycles on the properties of bio(nano)composites. For this series, the PLAPC blend was used as a reference. Compared with the pure PLA, the blend exhibited lower stiffness and strength and the same elongation at break. Further addition of NCC increased tensile stiffness and flexural elongation at 2 wt% addition, but decreased strength and elongation at break. Adding 5 wt% NCC decreased stiffness, strength, and elongation. An additional compounding cycle with 1 wt% NCC addition increased tensile stiffness and strength and maintained elongation at break, increased flexural strength, and decreased flexural stiffness and strength. An additional compounding cycle with 2 wt% NCC additive lowered tensile stiffness while maintaining tensile strength and elongation at break, lowered flexural stiffness and strength, and increased flexural strength. An additional compounding cycle with 5 wt% NCC addition reduced stiffness, strength, and elongation. Elongation was lower than for the PLA reference. It can be concluded that the reaction compounding conditions ensure good interfacial interactions between the thermoplastic matrix and the NCC and ensure good dispersion of the NCC in the thermoplastic matrix at NCC concentrations below 5 wt%. Comparing the first and the second reactive compounding cycles, it can be concluded that with one additional cycle of the reactive compounding with 2 wt% and 5 wt% NCC addition, degradation of PLA and possibly also of NCC already occurs, as evidenced by a decrease in tensile strength and elongation, which in the case of the PLA matrix is a good indicator of the onset of degradation of the PLA matrix, while the lower stiffness is an indicator of the onset of degradation of NCC. The degradation is most likely due to the high temperatures during reactive compounding and injection molding. It is more pronounced at higher NCC content, indicating simultaneous partial degradation of both the thermoplastic matrix and NCC. Reactive compounding of PLA and PC in the presence of a combination of two compatibilizers and a filler provides good miscibility of PLA and PC while ensuring good interfacial interactions and dispersion of NCC in the thermoplastic matrix at NCC concentrations below 2 wt%.

The highest tensile stiffness and strength were obtained for PLA15PBAT 5H with 2.59 GPa and 50.4 MPa, lower than the PLA reference (3.17 GPa and 71.8 MPa). The highest elongation at break was obtained for PLA20PBAT 1NCC with 39.9%, much higher than the pure PLA reference (4.8%). If good thermal stability of the bio(nano) composite is also required, then PLAPC 1NCC with a tensile stiffness of 2.37 GPa, strength of 40.5 MPa, and elongation at break of 9.6% would be the best choice.

3.2 Thermomechanical properties

The results of the dynamic mechanical evaluation are shown in **Figures 2–9**. The results for the PLA sample are used as reference values for all other modifications by reactive compounding and for the evaluation of the usability of the reactive compounding.

3.2.1 Results of the first-reactive compounding series

When hops were added to the blend of PLA and PBAT, the increasing amount of hops lowered the storage modulus in the glass transition region and allowed cold crystallization at lower temperatures. Simultaneously reducing the amount of hops and increasing the PBAT content further lowered the storage modulus from room temperature to the glass transition region. The onset of cold crystallization was comparable, and the height of the storage modulus was lower than that of the PLA15PBAT sample. The height of the peak of the loss factor at the glass transition of PLA in biocomposites decreased with increasing PBAT content. The position of the peak decreased with increasing hop content, and it also decreased with increasing PBAT content.

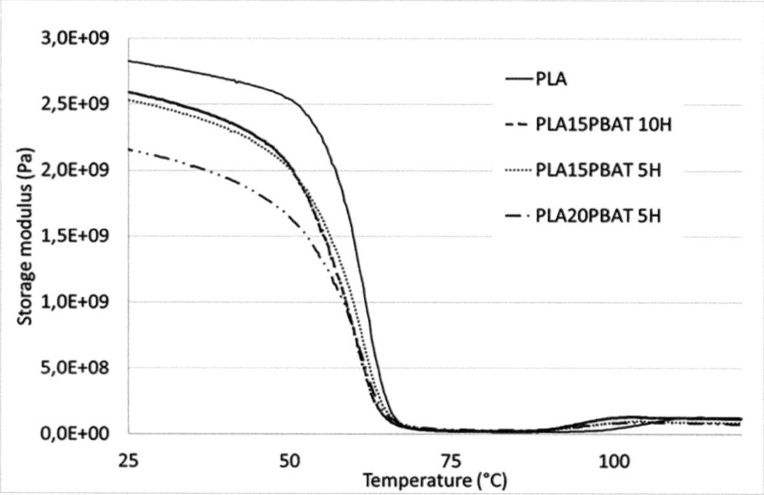

Figure 2.
Summarized results of storage modulus for the first series of the samples.

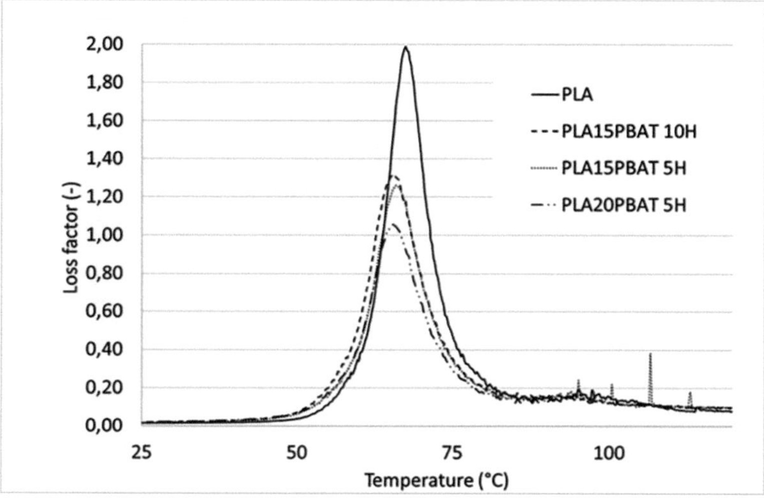

Figure 3.
Summarized results of loss factor for the first series of the samples.

Compared with the pure PLA reference material, all PLAPBAT biocomposites with hops had a lower storage modulus and also a lower glass transition temperature. The onset of cold crystallization in the biocomposites indicated good interfacial adhesion between the thermoplastic matrix and hops. PLA and PBAT, as well as homogenized hops, were successfully blended into biocomposites by reactive compounding.

3.2.2 Results of the second reactive compounding series

The addition of NCC was compared with the mixture of PLA/PBAT as a reference for the second series of samples. The addition of 1 wt% NCC reduced the storage modulus at the beginning of the glass transition and in other regions was comparable

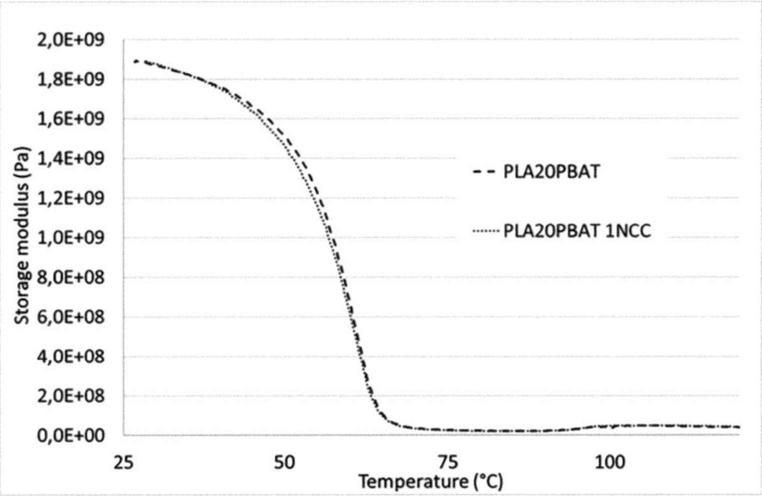

Figure 4.
Summarized results of storage modulus for the second series of the samples.

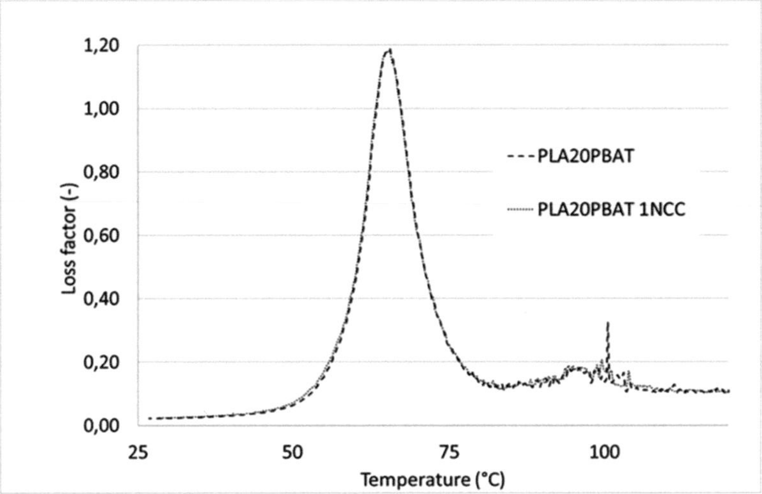

Figure 5.
Summarized results of loss factor for the second series of the samples.

with the reference PLA20PBAT. Compared with the sample from the first series PLA20PBAT 5H, the storage modulus of the sample PLA20PBAT 1NCC was lower due to the higher TPU content. The same results are shown in the dissipation factor, where the dissipation factor of the sample PLA20PBAT 1 NCC at the beginning of the glass transition was slightly higher than that of the reference PLA20PBAT. In addition, the onset of cold crystallization is seen slightly earlier and the peak is higher. The reactive compounding allowed good surface interactions between the thermoplastic matrix and the NCC, homogeneous dispersion of the NCC in the matrix, and good mixing of PLA and PBAT.

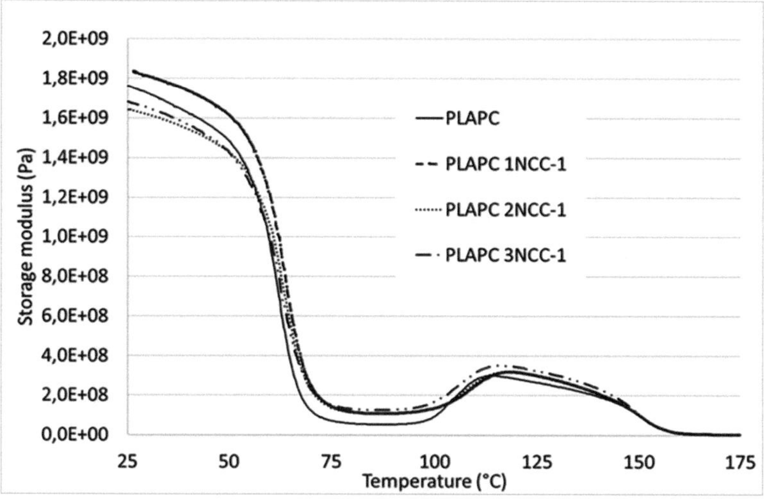

Figure 6.
Summarized results of storage modulus for the first compounding cycle.

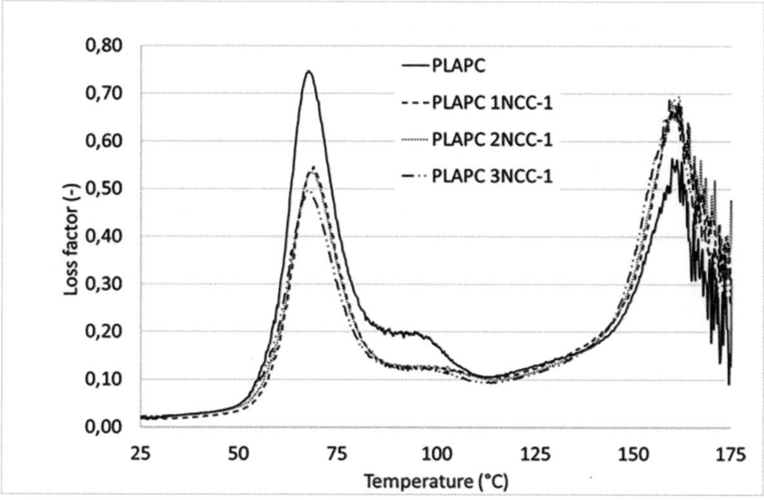

Figure 7.
Summarized results of loss factor for the first compounding cycle.

3.2.3 Results of the third-reactive compounding series

Storage modulus curves showed the first drop in glass transition temperature for PLA. The drop is more significant for PLA-based blends compared to PLA-based blends with NCC. The storage modulus was higher in this range (75–100°C) with higher NCC content. Above 100°C, the storage modulus increased due to cold crystallization of the material. The lowest peak for the cold crystallization temperature was for sample PLAPC (114°C), and the highest peak was for sample PLAPC 1NCC-2 (119°C). For all samples with NCC addition, the peak for cold crystallization temperature in the second compounding cycle was at a higher temperature than in the

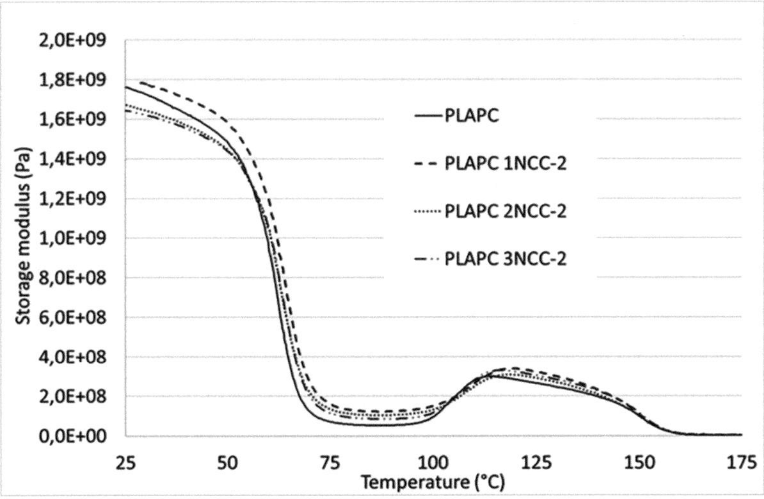

Figure 8.
Summarized results of storage modulus for the second compounding cycle.

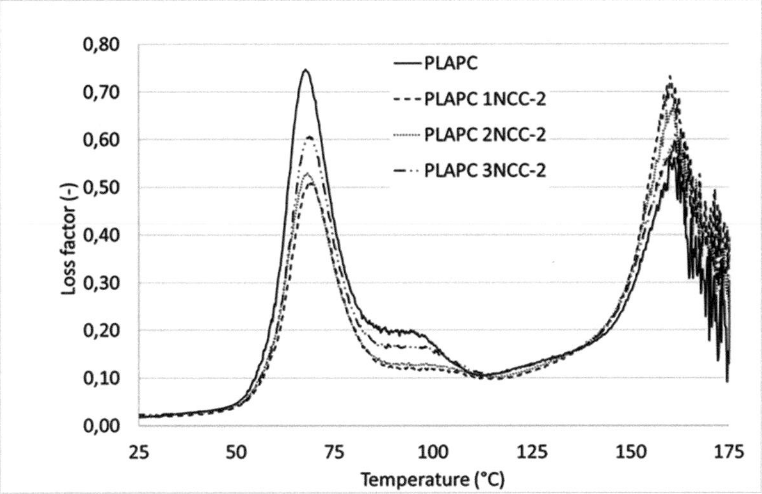

Figure 9.
Summarized results of loss factor for the second compounding cycle.

first compounding cycle. The addition of NCC to PLA-based compounds inhibited the cold crystallization of PLA. At low NCC loading (1 wt%), the NCC acted as a reinforcement for the PLA-based blend; at higher loadings (2 wt% and 5 wt%), the stiffness of the nanocomposites decreased below the stiffness of the neat matrix. Despite the inhibitory effect of NCC on cold crystallization, the NCC prevented softening of the PLA-based matrix after the glass transition temperature of PLA. The dissipation factor curve shows two sharp peaks at 69°C and 160°C (**Figures 4** and **5**) for PLA and PC matrices, respectively. The height of the first peak is the lowest for sample PLAPC 1NCC and the highest for sample PLAPC. The height of the second peak is the lowest for sample PLAPC and the highest for sample PLAPC 1NCC. In the second compounding cycle, the heights of the peaks were higher, indicating the beginning of the degradation of the matrix. The most elastic behavior is exhibited by sample PLAPC 1NCC after the first compounding cycle. The good surface interaction of NCC with the matrix due to the compatibilizer in sample PLAPC 1NCC increased the storage modulus maintained the storage modulus at a high level and decreased the height of the peak of the loss factor for PLA. The higher peak of dissipation factor for sample PC is due to the highest cold crystallization temperature for sample PLAPC 1NCC, and thus the overlap of the glass transition temperature of PC and the onset of melting of PLA.

The highest tensile stiffness and strength were obtained for PLA15PBAT 5H with 2.59 GPa and 50.4 MPa, which is lower than the PLA reference (3.17 GPa and 71.8 MPa).

The highest storage modulus up to the glass transition zone was achieved by the PLAPBAT samples with the addition of hops. The highest temperature stability was achieved with sample PLAPC 1NCC. All DMA results show that reactive extrusion is a suitable processing technology for bio(nano)composites even without surface modification of natural fibers.

3.3 Thermal properties

The results of the DSC evaluation are shown in **Table 5** and **Figures 10** and **11**. The results for the PLA sample were used as reference values for all other modifications by reactive compounding and for the evaluation of the usability of the reactive compounding.

3.3.1 Results of the first-reactive compounding series

When hops were added to the blend of PLA and PBAT, the increasing amount of hops lowered the cold crystallization temperature and melting temperature

Sample	T_g (°C)	ΔC_p (J/gK)	T_{cc} (°C)	ΔH_{cc} (J/g)	T_m (°C)	ΔH_m (J/g)	Diff. ΔH_m (J/g)
PLA	59.7	0.24	118.5	13.32	152.5	13.47	0.15
PLA15PBAT 10H	58.5	0.14	126.4	7.48	153.7	8.07	0.59
PLA15PBAT 5H	59.5	0.19	129.4	4.75	154.0	5.05	0.30
PLA20PBAT 5H	59.9	0.16	131.6	1.82	154.3	2.74	0.92
PLA20PBAT	60.5	0.33	125.8	4.01	150.8	5.82	1.81
PLA20PBAT 1NCC	60.3	0.36	139.1	0.29	150.9	0.60	0.31
PLAPC	59.8	0.11	123.6	5.35	152.0	5.37	0.02
PLAPC 1NCC-1	60.8	0.09	133.3	0.32	152.2	0.37	0.05
PLAPC 1NCC-2	60.6	0.09	128.3	0.04	152.2	0.33	0.29
PLAPC 2NCC-1	61.1	0.08	133.4	0.13	152.4	0.31	0.18
PLAPC 2NCC-2	60.7	0.09	131.6	0.18	152.8	0.52	0.34
PLAPC 5NCC-1	61.1	0.06	130.4	0.24	153.2	0.48	0.24
PLAPC 5NCC-2	60.1	0.08	128.4	1.28	152.9	1.67	0.39

Table 5.
Summarized results from the second heating from DSC tests.

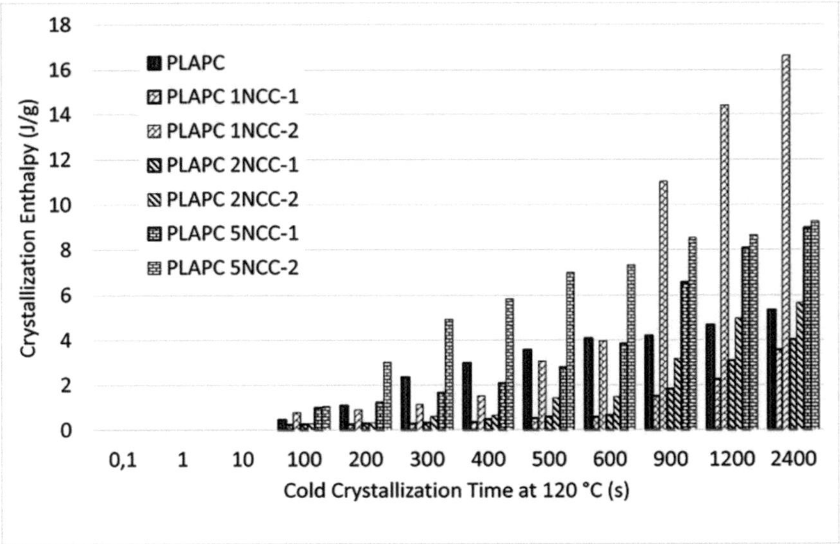

Figure 10.
Summarized results of crystallization enthalpy of the samples after aging at 90°C for 100 s and various cold crystallization times at 120°C.

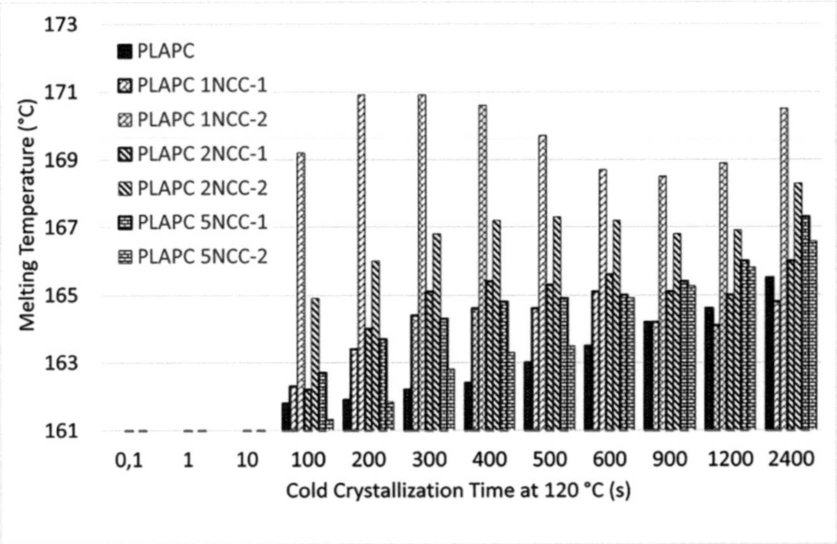

Figure 11.
Summarized results of melting temperature of the samples after aging at 90°C for 100 s and various cold crystallization times at 120°C.

and improved the crystallinity. The glass transition temperature was not affected. Simultaneously reducing the hops content and increasing the PBAT content increased the cold crystallization temperature, melting temperature, and crystallinity. Increasing the PBAT content increased the cold crystallization temperature and crystallinity. Compared with the pure PLA reference, the cold crystallization temperature, melting temperature, and crystallinity were increased for biocomposites.

3.3.2 Results of the second-reactive compounding series

The addition of NCC was compared with the mixture of PLA/PBAT as a reference for the second series of samples. With the addition of 1 wt% NCC, the cold crystallization temperature increased and the crystallinity was reduced.

3.3.3 Results of the third-reactive compounding series

At the first compounding cycle, the cold crystallization temperature, melting temperature, and crystallinity were increased for all bio(nano)composite samples compared with the PLAPC reference. During the second compounding cycle, the cold crystallization temperatures were decreased and the crystallinity was increased.

Higher crystallinity indicates good homogeneity of the bio(nano)composites.

The crystallization kinetics were characterized with flash DSC for the third series of samples (**Figures 10** and **11**). The onset of formation of the crystal units was after aging at 90°C for 100 s and crystallization at 120°C after 100 s for all samples. The shorter time is not sufficient for the formation of the crystal moieties formation. Higher NCC loading promoted the crystal moieties formation as well as the second compounding cycle. The fastest and largest increase in crystal moieties was for sample PLAPC 5NCC-2 up to the cold crystallization time 600 s, and then for sample PLAPC 1NCC-2. Therefore, the conclusion can be made that NCC inhibited cold crystallization at shorter times and enhanced cold crystallization at longer times at elevated temperatures. The mobility of PLA chains at elevated temperatures reached a threshold for the formation of crystal units after 600 s at 1 wt% and 2 wt% NCC loading. At 5 wt% NCC loading, 100 s is sufficient due to the higher amount of NCC particles in the matrix. From the results, we can also conclude that agglomeration of NCC occurs to a smaller extent and increases with increasing NCC content.

3.4 Impact properties

The results of the toughness evaluation are shown in **Figure 12**. The results for the PLA sample are used as reference values for all other modifications by reactive compounding and for the evaluation of the usability of the reactive compounding.

3.4.1 Results of the first-reactive compounding series

Higher hops content and lower PBAT content in biocomposites PLAPBAT/hops lowered impact strength. On the contrary, a lower hops loading and a higher PBAT loading led to a larger scatter of the measurement results.

3.4.2 Results of the second-reactive compounding series

The addition of NCC was compared with the mixture of PLA/PBAT as a reference for the second series of samples. Both the reference and the PLA20PBAT 1NCC sample showed excellent impact resistance, as they did not break during the impact test.

3.4.3 Results of the third-reactive compounding series

The addition of NCC to the PLAPC mixture improved the toughness of the bio(nano)composites. The exception is the sample PLAPC 5NCC-2 due to the

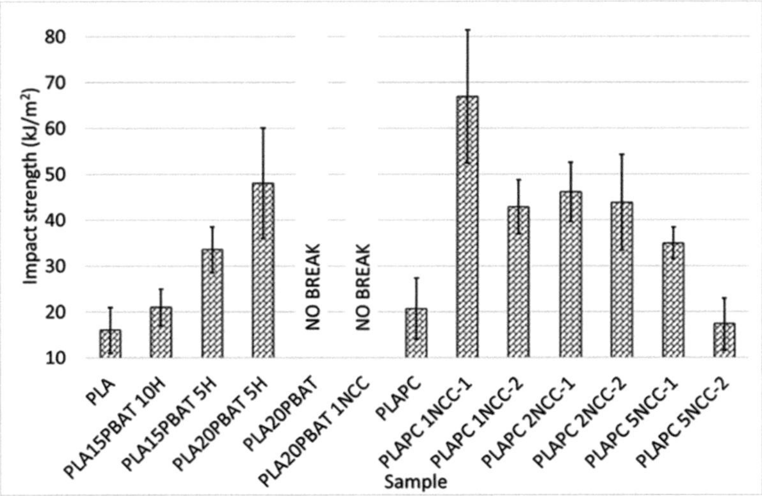

Figure 12.
Summarized results of impact strength.

degradation of PLA and NCC. The best toughness was obtained for sample PLAPC 1NCC-1. Higher NCC loading decreased the toughness as well as the second compounding cycle.

Sample PLA20PBAT 1NCC showed the best toughness, followed by sample PLAPC 1NCC and sample PLA20PBAT 5H. It is obvious that NCC has successfully improved the toughness of bio(nano)composites through the appropriate processing technology—reactive compounding.

4. Conclusion

Reactive compounding was used for bio(nano)composites with PBAT and PC in addition to the main PLA matrix together with appropriate compatibilizers and (processing) additives. The adequacy of the reactive compounding was evaluated by characterizing the mechanical, thermomechanical, and thermal properties, as well as toughness.

The evaluation of mechanical properties showed that novel properties were achieved by the addition of NCC. The blend was able to achieve either high toughness with the addition of PBAT or high-temperature stability with the addition of PC. The prepared bio(nano)composites showed good miscibility of PLA and PBAT or PLA and PC and good surface interaction between the thermoplastic matrix and the natural fibers, although the surface of the natural fibers was not modified. Furthermore, the flash DSC results showed an altered morphology behavior of the PLAPC 1NCC-2 bio(nano)composite. Longer residence time at elevated temperature accelerates crystallization as a result of the degradation of PLA and NCC due to shorter PLA chains and smaller NCC particles, which acts as nuclei for the initiation of heterogeneous crystallization of PLA. At the same time, we can observe that the NCC is well distributed in the thermoplastic matrix due to the increasing crystallinity. For the PLA/PBAT blends, good miscibility was achieved with the proper processing parameters and by using appropriate chain extenders. Good surface interaction between the thermoplastic matrix and the natural fibers was achieved with the proper compatibilizers and loading. The adequacy of the reactive compounding was evaluated by the simultaneous increase in stiffness and elongation at break in the tensile test, the change in storage modulus and loss factor in DMA, the change in cold crystallization temperature and crystallinity in DSC, and the increase in impact strength. The prepared bio(nano)composites showed tougher behavior while maintaining high stiffness and strength. The addition of NCC also affected the morphology of the bio(nano)composites, which can be controlled by the processing parameters. The second reactive compounding cycle at 1 wt% NCC loading showed that recycling of the novel bio(nano)composites can also be performed without much influence on the properties of the recycled products. The present work shows that the existing polymer processing equipment is suitable for the production of bio(nano)composites and their recycling. Sustainable design was the guiding principle for conducting the research with surface-unmodified natural fibers to avoid the use of chemicals and thus minimize the impact of bio(nano)composites on the environment.

Reactive compounding is a suitable processing technology for bio(nano)composites, even if the surface of natural fibers is not modified, to achieve novel properties of PLA-based blends with natural fibers (preferably NCC). The desired properties can be developed by suitable compatibilizers and processing additives during reactive compounding. To describe the dependence on the amount of added NCC in bio(nano)composites, the addition of less than 1 wt% NCC in PLA-based blends bio(nano)composites should be investigated in further research.

Acknowledgements

We would like to thank Navitas for providing nanocrystalline cellulose.

Author details

Silvester Bolka* and Blaž Nardin
Faculty of Polymer Technology, Slovenj Gradec, Slovenia

*Address all correspondence to: silvester.bolka@ftpo.eu

References

[1] Ma P. Tailoring the Properties of Bio-Based and Biocompostable Polymer Blends. Eindhoven, Netherlands: Eindhoven University of Technology; 2011

[2] Zeng JB, Li KA, Du AK. Compatibilization strategies in poly(lactic acid)-based blends. RSC Advances. 2015;**5**:32546-32565. DOI: 10.1039/c5ra01655j

[3] Murariu M, Dubois P. PLA composites: From production to properties. Advanced Drug Delivery Reviews. 2016;**107**:17-46. DOI: 10.1016/j.addr.2016.04.003

[4] Tolinski M. Additives for Polyolefins: Getting the Most out of Polypropylene, Polyethylene and TPO. Oxford, William Andrew; 2009

[5] Dluzneski PR. Peroxide vulcanization of elastomers. Rubber Chemistry and Technology. 2001;**74**:451-492. DOI: 10.5254/1.3547647

[6] Kruželák J, Sýkora R, Hudec I. Vulcanization of rubber compounds with peroxide curing systems. Rubber Chemistry and Technology. 2017;**90**:60-88. DOI: 10.5254/rct.16.83758

[7] Södergård A, Niemi M, Selin JF, Näsman JH. Changes in peroxide melt-modified poly(L-lactide). Industrial and Engineering Chemistry Research. 1995;**34**:1203-1207. DOI: 10.1021/ie00043a024

[8] Takamura M, Nakamura T, Takahashi T, Koyama K. Effect of type of peroxide on cross-linking of poly(l-lactide). Polymer Degradation and Stability. 2008;**93**:1909-1916. DOI: 10.1016/j.polymdegradstab.2008.07.001

[9] Takamura M, Nakamura T, Kawaguchi S, Takahashi T, Koyama K. Molecular characterization and crystallization behavior of peroxide-induced slightly crosslinked poly(L-lactide) during extrusion. Polymer Journal. 2010;**42**:600-608. DOI: 10.1038/pj.2010.42

[10] Carlson D, Dubois P, Nie L, Narayan R. Free radical branching of polylactide by reactive extrusion. Polymer Engineering and Science. 1998;**38**:311-321. DOI: 10.1002/pen.10192

[11] Signori F, Boggioni A, Righetti MC, Rondán CE, Bronco S, Ciardelli F. Evidences of transesterification, chain branching and cross-linking in a biopolyester commercial blend upon reaction with dicumyl peroxide in the melt. Macromolecular Materials and Engineering. 2015;**300**:153-160. DOI: 10.1002/mame.201400187

[12] Rytlewski P, Zenkiewicz M, Malinowski R. Influence of dicumyl peroxide content on thermal and mechanical properties of polylactide. International Polymer Processing. 2011;**26**:580-586. DOI: 10.3139/217.2521

[13] Formela K, Zedler A, Hejna A. Tercjak, reactive extrusion of bio-based polymer blends and composites–current trends and future developments. Express Polymer Letters. 2018;**12**:24-57. DOI: 10.3144/expresspolymlett.2018.4

[14] Li X, Yan X, Yang J, Pan H, Gao G, Zhang H, et al. Improvement of compatibility and mechanical properties of the poly(lactic acid)/poly(butylene adipate-co-terephthalate) blends and films by reactive extrusion with chain extender. Polymer Engineering and Science. 2018;**58**:1868-1878. DOI: 10.1002/pen.24795

[15] Quiles-Carrillo L, Montanes N, Lagaron JM, Balart R, Torres-Giner S. In situ Compatibilization of biopolymer ternary blends by reactive extrusion with low-functionality epoxy-based styrene–acrylic oligomer. Journal of Polymers and the Environment. 2019;**27**:84-96. DOI: 10.1007/s10924-018-1324-2

[16] Farias da Silva JM, Soares BG. Epoxidized cardanol-based prepolymer as promising biobased compatibilizing agent for PLA/PBAT blends. Polymer Testing. 2021;**93**:106889. DOI: 10.1016/j.polymertesting.2020.106889

[17] Andrzejewski J. Development of toughened flax fiber reinforced composites. modification of poly(lactic acid)/poly(butylene Adipate-co-terephthalate) blends by reactive extrusion process. Materials. 2021;**14**(6):1523

[18] Singha NR, Mahapatra M, Karmakar M, Chattopadhyay PK. Processing, characterization and application of natural rubber based environmentally friendly polymer composites. In: Inamuddin Thomas S, Kumar Mishra R, Asiri A, editors. Sustainable Polymer Composites and Nanocomposites. Cham: Springer; 2019. DOI: 10.1007/978-3-030-05399-4_29

[19] Martínez-Sanz M, Lopez-Rubio A, Lagaron JM. Optimization of the dispersion of unmodified bacterial cellulose nanowhiskers into polylactide via melt compounding to significantly enhance barrier and mechanical properties. Biomacromolecules. 2012;**13**:3887-3899. DOI: 10.1021/bm301430j

[20] Zhao X, Hu H, Wang X, Yu X, Zhou W, Peng S. Super tough poly(lactic acid) blends: A comprehensive review. RSC Advances. 2020;**10**:13316-13368. DOI: 10.1039/d0ra01801e

[21] Hashima K, Nishitsuji S, Inoue T. Structure-properties of super-tough PLA alloy with excellent heat resistance. Polymer (Guildf). 2010;**51**:3934-3939. DOI: 10.1016/j.polymer.2010.06.045

Chapter 4

The Porosity of Nanofiber Layers

Sedigheh Aghayari

Abstract

Nanofiber layers have recently received lots of attention. These layers can be produced in various methods, but the most common is electrospinning. Therefore, this chapter focuses on the nanofiber layers from electrospinning. The porosity of nanofiber layers is a critical property. Several methods can be used to measure this value. Also, there are numerous methods for controlling and changing it. The porosity is an essential property for the application of nanofiber layers. Each application requires a unique set of porosities. As a result, measuring and controlling the porosity with high precision is critical for applying nanofiber layers. This chapter concentrated on porosity measurement and control methods and the importance of porosity in applications.

Keywords: nanofiber, porosity, electrospinning, nanofiber layers, porosity measurement methods

1. Introduction

In recent years, multifunctional properties for nanofibers derived from polymers, metal composites, and metal oxides have been expressed. Additionally, surface-modified nanofibers are simple and inexpensive to manufacture. It is due to the unique properties of nanofibers, which include high tensile strength, high specific surface area (surface area per unit mass), and porosity [1, 2]. The length-to-diameter ratio of nanofibers is high [3]. As a result, nanofiber properties are critical for high-performance filters, absorbent textiles, medical textiles, drug release, and many other applications [1, 4].

Electrospinning is a continuous method for producing nanofibers with diameters ranging from micrometers to nanometers. Layers with a high specific surface area, high porosity, and good mechanical properties can be produced using this method [5].

Electrospinning is a popular method for preparing scaffolds. Various electrospun nanofiber patterns are used to prepare layers with medical applications ranging from artificial skin to endocrine organs and from the nervous system to cardiovascular applications [6].

The porosity of electrospun layers varies depending on their application. In some applications, porosity is required less than usual, while in others, it is required more than usual. Therefore, there are various methods for achieving sufficient porosity.

The only way to prepare large-scale nanofibers is through electrospinning. The reason for this is the ease of control, high speed, low solution consumption, control of diameter and pores and fibers alignment, ease of the process, low cost, simple and

IntechOpen

reproducible fiber production process, and technological advances [7]. This process can use a wide range of polymers to obtain polymer fibers in the submicron range, which is difficult to achieve with traditional spinning methods [8]. Electrospinning is affected by various environmental and solubility parameters and processes. Two environmental parameters include temperature and relative humidity. Also, concentration, conductivity, molecular weight, and viscosity are the solubility parameters. The process parameter includes feeding rate, voltage, and needle distance to the collector. The diameter of fibers decreases with increasing temperature and increases with increasing humidity [6, 7]. The electrospun fibers had diameters ranging from 10 nm to 100 μm. Polymer solutions are most commonly used for electrospinning. However, in some cases, polymeric melts with higher direct current voltages can also produce fibers with diameters less than micrometers [6].

Electrospinning is an electrohydrodynamic process in which the liquid droplet is affected by electricity to make a jet (accelerated flow of liquid). It is then stretched and bent to form fibers or nanofibers with the main components of the high voltage supply, a power pump (in the form of a syringe), a spinner (typically a special subcutaneous injection needle with a tipless head), and a conductive collector. Power can be supplied in either a direct or alternating current. Surface tension causes the liquid to be removed from the filament during electrospinning, resulting in a hanging drop. As soon as the droplet is electrified, the repulsive force between the same surface loads reshapes it as a Taylor cone, ejecting a loaded jet. The jet is first drawn in a line (this occurs due to polymer tying in a polymer solution and prevents the jet from turning into droplets [9]), and then it undergoes rapid whipping movements due to bending instability (this occurs due to high surface load density [9]). The jet quickly solidifies as it is drawn to finer diameters, resulting in solid fiber deposition on the surface of the collector attached to the ground. Typically, the electrospinning process is divided into four consecutive steps:

1. Charging liquid drops and formation of Taylor cones or cone-shaped jets;

2. Drawing the charged jet along a straight line;

3. Jet thinning in the presence of an electric field and increasing electrical bending instability (also known as whipping instability); and

4. Freezing and collecting jets in the form of fibers or solid forks on the grounded collector [10].

Because of their fine diameter, these layers have a high specific surface area, high porosity, and small pore sizes [11, 12]. If the fibers also have pores, the porosity and the specific surface area will also increase [10]. Femme and his colleagues demonstrated that increasing the diameter of the fibers increased the pore size [13]. This increase will cause a reduction in the specific surface area. Therefore, if high specific surface area, fine diameters, and large pores are required in applications, the pore size must be de-linked from the fiber diameter. Both theoretical models and experimental studies have shown that nanofiber diameter strongly affects the pore diameter with smaller nanofiber diameters resulting in smaller pores [11] (which will be discussed in the future). Higher pressure is also required for any fluid to enter the layer (higher surface tension) [8].

Smart layers, filtration membranes, fuel cells, batteries, wound dressings, sensors, catalysts, energy storage cells, electronics, and spintronics use electrospinning [1].

Drug delivery, tissue engineering, and protective textiles benefit from porosity and high specific surface area [4]. Because of their high porosity, good mechanical properties, and high-water permeability, electrospun membranes will be practical for air filtration and water purification [5]. Electrospun layers are used in multi-structural thin films, ultrafiltration, nanofiltration, reverse osmosis, and distillation membranes as porous protective layers. Today, distillation membranes are highly regarded. Multi-structural thin film membranes have three layers:

1. An upper layer that is very thin for filtration,

2. A middle layer that is a porous protective layer, and

3. The underlying layer that is a nonwoven textile [7].

For automotive air filters, nanofibers are coated on a standard filter environment to allow high efficiency and long filter life, which occurs with the lowest increase in pressure drop [12].

Electrospun nanofiber layers have a high specific surface area, controllable porosity, interconnective pores, microscale interstitial distance, and flexibility because of their various sizes and morphologies. Due to these advantages, they are appealing in applications [14]. However, because of electrospinning disadvantages, such as high voltage sources, it is necessary to use toxic organic solvents and low production rates [15]. Synthesis conditions (like high humidity) can affect nanofibers' morphology, such as mesoporous fibers [16], resulting in a higher specific surface area of nanofibers that can improve their properties.

The amount of air, gas, or vacuum in a solid material is often expressed as the percentage of the nonsolid portion volume divided by the total volume (total solid and nonsolid volume) of a unit of matter [17]. Porosity is the fluid volume or space in the filter media to the total volume of the filter ratio. It has no unit, and its value can range from zero to one. One of the essential parameters in the design and operation of filters is porosity. Nanofiber layers have made nanofiber coatings the most important candidates for high-performance filters due to their porosity and sufficient surface area [12]. The membrane's porosity, pore size distribution, and bending make it simple to pass steam through and collect steam as a filter outlet [9]. Previous research has shown that changing fibers' sedimentation rate can control the nanofiber layer's thickness and porosity. The pore space in the layer is related to the total porosity in the electrospinning layers [12].

Nonwoven materials have a pore structure, which is critical in their application. There can be three types of pores in a matter [18]. Closed pores are inaccessible, they also restrict the blind pores within the material and prevent fluid from passing through. Open pores are outwards and allow fluid to pass through, many nonwoven textile applications benefit from open pores. Their main characteristics are the largest pore diameter, pore distribution, high specific surface area, and gas permeability liquid passage. Through pores are pores which are in the entire thickness of the layer and the fluid can enter and then leave the layer through them.

1.1 Methods of changing the porosity of nanofibrous webs

Porosity can be created on the fibers and referred to as porous nanofibers. However, porosity can also be created between the fibers. The porosity that forms on the fibers can be reduced using the following methods:

1. Phase separation between polymer and solvent (breath figure) [10, 19];

2. Phase separation between polymer and nonsolvent [20];

3. Phase separation of the liquid section in polymer, nonsolvent, and solvent systems [21];

4. Collecting the nanofibers in a nonsolvent bath [21];

5. Selective removal of one phase of nanofiber [10]; and

6. Thermal degradation of one phase [22].

Many methods are used to increase the porosity of layers in which the porous scaffold structure is naturally placed together by adding the macroporous structure. This process is accomplished by aligning crystalline structures like ice or salt crystals parallel to the electrospinning. However, these methods increase cell penetration in the electrospun scaffold while preventing the actual pore effect defined by single fibers [10].

Fiber collection on a rotating axis is a technique for modifying porosity and pore size independently of fiber diameter. Porosity decreases as axis speed increases due to increased layer density. At the same time, the diameter of the fibers decreases slightly [10].

1.2 Performing a final treatment

In this case, one of the components must be water soluble while the other is not. The pore size of a fiber changes when it is dissolved in water. It has been demonstrated that dissolving a component in water causes the pore size to increase exponentially. The effect on cell penetration is minimal in this condition, but the significant fiber removal from the layer affects the mechanical properties. Another person demonstrated that this work doubled the size of the pores and completed penetration into the cell, whereas previous work only affected penetration into the surface cell [23].

Of course, numerous methods exist for modifying pore size and porosity independent of the fiber diameter [24]. According to studies, the diameter of the fibers strongly influences the diameter of the pores and the porosity. The smaller diameter of the fibers, the smaller the pores. However, for some applications, it is appealing to combine the increased specific surface area provided by fine fibers with large pores for cell or fluid transport. For this purpose, the increase in pore size should be made independent of the fiber diameter.

To this end, two different polymers can be purposefully mixed during electrospinning. Then, one of these polymers is selectively dissolved, increasing the void volume and the pore size. The layer is created for this purpose by electrospinning two polymers from two unique devices (side-by-side arrangement). This configuration can result in high output.

The resulting fibers are layered on top of one another but are not intentionally mixed. The tendency of the resulting fibers to prevent material mixing in the case of side-by-side charged jets can be explained simply by the electrostatic repulsion of materials with the same charge. For this purpose, one can use the core/shell or side-by-side arrangement with a spinning machine and various other arrangements. In this case, the spinning conditions for both polymers cannot be independent.

Moreover, both polymers must be solvent-soluble, and the applied voltage must be the same [11]. A simple arrangement can spin two polymers with different solvents and applied voltages. Each material has its unique spinneret and voltage supply. Because the charged droplets are evenly spaced on the grounded collector, no electric field is applied to the rotating collector, and no electric field is observed. The collector rotates fast enough to mix the different fibers. However, it cannot make arranged or non-isotropic layers. Two electrospinning units are placed facing each other with a collector. Each machine's electrospinning parameters are controlled independently. The fibers are mixed by the collector connected to the ground to form a non-woven layer [11].

By performing an end treatment on the two-dimensional (2D) layer, three-dimensional foam can be obtained. A suspension of short fibers is frequently obtained by shearing and homogenizing the layer in a liquid (usually water or ethanol depending on the solubility of the nanofibers). After drying and cooling, a super elastic three-dimensional (3D) super light nano fiber foam is obtained. This foam has a low density and extremely high porosity. A cross-linking agent is required to increase the interaction between these short fibers [10]. Fiber collection engineering is a straightforward method for directly assembling nanofibers in three dimensions by engineering the aggregate [10]. Another method for creating 3D foams is to immerse the two-dimensional layer in a sodium borohydride aqueous solution. When this material is rapidly hydrolyzed in water, hydrogen gas passes through the 2D layer under applied pressure, separating the nanofibers and forming a 3D structure [25]. Furthermore, dry ice causes the same process by producing carbon dioxide; however, this method does not require water usage. The porosity is significantly high with this method [10].

The last three methods are for creating a 3D sponge with an irregular structure that lacks topographical cues. The materials required for these methods are exceptional. To overcome this limitation, the gas sponge method, which employs borohydride to create a very regular 3D sponge, employs a non-axially regularized layer [10].

Scaffold porosity is affected by electrospinning conditions, and fibers with varying fiber diameters made from various polymers can be produced in a controlled and reproducible manner [26–28].

Applying photoinduced thiol-ene cross-linking reactions to the electrospun layer is a versatile and efficient method for tuning the porosity of the nanofiber's web. Aside from preventing the polymer cold flow and freezing the structure obtained by electrospinning, the photocuring step finely controls the morphology of the nanofiber layers [29].

Electrospun porous nanofibers loaded with photocatalytic particles can increase the particle contact area with light, shorten the electron transfer path, and improve photocatalytic activity, which has applications in pollutant degradation [24].

Porosity can be calculated using the formula with the nanofibers' pore size and diameter [5].

2. Application of reducing and increasing layer porosity

An example of electrospinning layers with typical porosity and low porosity obtained by electrospinning with two pumps is examined. A wide lumen with no leakage necessitates blood flow in the lumen of a scaffold with no vessel leakage. An important parameter to consider is the size of the pore. If the pore size is too small, it prevents penetration into the cell. On the other hand, if it is too large, blood leakage occurs. According to research, the performance of double-layered scaffolds inside and outside the body, with one layer with low porosity (62%) and another with high porosity (81%), is also placed in the conduit or external coating. A comparison of vascular scaffolds made of a single layer of nanofibers with high porosity, with two-layer scaffolds that significantly reduce blood leakage, reveals that a layer with low porosity is required. Also, when a multilayered vascular scaffold is made, leakage is possible. The blood level should be low and cell penetration should be done well. Cell infiltration from the connective tissue surrounding the scaffold was also concluded to be greater than flowing blood [10]. It is necessary to compactly place fibers with a very small diameter to achieve a layer with low porosity. In general, a two-pump electrospinning machine is required for this lumen. This is done to keep the two polymer solutions from mixing. The transfer is continuous in these layers, and the layers are not separated. Less electrospinning time is required to produce a layer with less porosity. The electrospinning conditions of each layer are different from the others [30]. Therefore, in some applications, a layer with low porosity and small pore size is required, whereas other common electrospun layers are suitable. Of course, in some applications, it is necessary to make an ultra-porous layer with large pore size and fine fibers to transport cells or fluid [11]. A layer with high porosity and good protection is required in some applications, such as wound dressing. Wound dressing permeability against oxygen, control of water vapor exit, and fluid passage occur due to high porosity in the layer of nanofibers, resulting in wound healing [17].

3. Porosity measurement methods

Electrospun polymeric nanofibers have broad applications, such as automobile air filters. High specific surface area, small pores, flexibility, and sufficient porosity are essential for improving the filter media performance, so measuring porosity is critical. Porosity affects several material properties, such as sorption capacity and mechanical, thermal, and electrical properties [31]. For example, porous materials can store and transport gas. Their porosity affects hydrate formation and gas-storage capacity [32]. Also, the quality of the dental composite depends on its degree of porosity [33, 34], so measuring the porosity by the most accurate method is very important. Porosity measurement methods such as the density method, mercury porosimetry, image analysis, and capillary flow porometry are relatively inaccurate methods, and all have disadvantages for measuring the porosity of nanofibers. Another way to measure porosity is by expanding gas, which works with helium gas. To date, accurate porosity measurement in nanofiber layers has been challenging.

3.1 Density method

The overall porosity is obtained following the conventional method per the following equation [8]:

$$Total\ Porosity = 1 - Layer\ Density \quad (1)$$

In this equation, the material density relates to the material from which the layer is obtained. Also, layer density is calculated by dividing weight by layer volume. This method is a good alternative when there is no access to the right equipment because it is convenient and fast. Samples should be prepared carefully, and dimension measurements should be accurate to achieve precise accuracy. A micrometer was used to obtain the samples' thickness, and the diameter of the samples was used to obtain the sample size [8, 35]. This method causes many errors in the actual porosity [36]. Of course, this method is suitable for nanofiber layers that cannot withstand high pressure [37].

3.2 Mercury method

This method is frequently used to characterize structures with nonwoven pores. Mercury is not very destructive to soils because the surface free energy between mercury and soil is more sizable than between gas and soil. Mercury does not enter the pores. However, it can enter with force. Pressure is needed to enter mercury into the pore, obtained by the pore diameter. The measurement of injectable pressure and intrusive volume obtains the diameter and volume of blind and open pores. In this method, mercury enters the pores with compulsion and pressure, the volume of mercury is penetrated, and the pressure is measured. This method only reveals the volume of pores and the diameter of blind and open pores. Here, the essential characteristics are no longer measured.

Additionally, this method requires high pressure, which can significantly damage the pore structure of nonwoven materials. Mercury is used in this method, which harms one's health and pollutes the environment. The pressure applied in mercury porosity is slightly higher than 0.5 to 60,000 pounds per square inch. Biological materials that can be compressed or degraded at high pressures must be analyzed with relatively quantitative pressures, or a correction for compression capability should be applied to experimental measurements. The pressure is obtained as per the following equation:

$$Pressure = 2\,/\,Pore\ Diameter \tag{2}$$

Open porosity (obtained by the mercury measurement method) is obtained from the following equation:

$$Open\ Porosity = Mercury\ Penetration - Total\ Volume\ Layer\ Volume \tag{3}$$

Finally, the closed porosity, which cannot be calculated by mercury porosity, is determined by the following equation [38]:

$$Close\ Porosity = Total\ Porosity - Open\ Porosity \tag{4}$$

This method obtains the average pore diameter, size distribution, and total volume fraction of pores [19, 39]. It should be noted that mercury is expensive and toxic.

However, it should be noted that this method is generally a liquid penetrating method that can be used with Vaseline. In this case, the layer is weighed and remains in Vaseline at room temperature with a mechanical stirrer for one day to allow the liquid to penetrate the volume of the layer. The surface of the samples was then dried and weighed again to obtain the weight of the penetrated Vaseline. Measurements are performed on five samples, and porosity is obtained from the following relationship:

$$Porosity = Total\ Vaseline\ Volume\ /\ Volume\ of\ layer \quad (5)$$

The Vaseline volume is derived from the division of primary and secondary Vaseline mass differences. The layer volume was obtained from the total Vaseline volume and the volume of fibers. Fibers' volume is derived by dividing the primary mass by the density of the polymer of the layer manufacturer [8].

3.3 Liquid intrusion method

The fluid intrusion method can also achieve open porosity. This method is derived from the method for calculating the pore diameter. First, the method for calculating pore size is investigated. This method provides the most detailed information about the structure of the pores within the layer and is appropriate for testing the polymeric nanofiber layer. The general rule is that a liquid with a lower surface energy than the sample with gas is used to fill the pores of the sample. A decrease in the system's free energy causes pores to fill spontaneously. Gas can transport liquid through pores, and the gas increases the free energy of the surface, allowing the free energy of the surface to be moved from the low-level free energy between the sample and the liquid to the high-level free energy between the sample and the gas.

If the sample is simply dropped into the liquid and removed, the porosity can be calculated using the fluid displacement method as shown below. After removing the saturated layer from the liquid, the residual liquid volume and open porosity are calculated using the following equation [38]:

$$Porosity = residue\ liquid\ volumeInitial\ Volume \quad (6)$$

However, according to some sources, this liquid should not cause the layer to accumulate and swell [40].

3.4 Scanning electron microscopy images

Various computer software programs analyze this method to determine porosity, particularly pore size. Samples of various sizes are used for statistical analysis. Finally, they shoot and analyze the porosity and pore size in layers using industrial shears. In short, isotropic incision information is obtained and reconstructed into two-dimensional images, which are then written and analyzed to create 3D photographs and to obtain few morphological details.

Surface porosity is calculated using scanning electron microscopy images by dividing the free area by the total area of the sample. For the fiber area, the average diameter is used [8].

$$Porosity = \left(Total\ sample\ area - total\ fiber\ area\ /\ total\ sample\ area\right) * 100 \quad (7)$$

However, in this case, only surface pores are considered. Regardless of the layer with higher fiber density, the porosity values in this method are comparable to those in the liquid intrusion method [8].

3.5 Gas expansion

In this method, the solid density is measured using a volume of 150 cc. The volume of nanofibers can be calculated using a density meter and porosity based on the ratio of empty space volume to total volume. The total volume of the image thickness was determined by scanning electron microscopy (SEM). The use of a density meter to calculate pore volume is not novel, but it is new for determining the porosity of nanofiber layers via filtration. The density meter measures volume; however, the density can be calculated using the sample's mass. The helium gas density meter is in accordance with the gas law. In this device, the volume is measured by the amount of fluid pressure change, resulting in the displacement of the sample in a constant volume. This method begins with one chamber (chamber 1) that is initially empty and has atmospheric pressure. As a result, the first chamber's confusing pressure is zero. Chamber 1 have pressures and volumes of P1 and V1. The pressure and volume of the second chamber are P2 and V2, respectively. In the initial state, the valve is open and the two chambers are at equilibrium pressure P3, so the equilibrium equation is written as follows:

$$P3(V1+V2)=P2V2 \tag{8}$$

Next, the sample is placed in chamber 1, and the pressure of the second chamber is set to P4.

$$P5((V1-Vf)+V2)=P4V2 \tag{9}$$

$$VF=V1-P3V1(P5-P4)P5(P3-P2) \tag{10}$$

As a result, the volume of VF nanofibers can be obtained by setting up the density meter twice and measuring the pressure of P2, P3, P4, and P5 with the specified volume of V1. Also, the volume of fibers can be calculated as porosity with the following equation:

$$Porosity=1-(Total\ Fiber\ Volume\ /\ Total\ Web\ Volume) \tag{11}$$

This method has less porosity than the traditional method and has advantages over the others mentioned previously. The density of polyamide 6 chips was measured to control the accuracy of this method.

1. The minimum damage to the fibers,

2. Chemical entry,

3. In this method, meso- and micropores are both measured, and

4. Using another medium is not required due to applying static pressure on the layer [12].

The nanofiber layer's porosity is an important property, and any application requires specific porosity [41]. All of the porosity measurement methods that were discussed had drawbacks. Methods based on liquids had more difficulty creating the sample, resulting in changes in structure and measurement errors. In addition, non-wetting liquids can damage the structure due to the need for pressure, which can change the structure and even damage the layer. The density method also has errors due to the need to measure dimensions, and a scale with appropriate accuracy is required. Of course, SEM images can be used to accurately measure dimensions. However, there are some issues to consider. The thickness of the layers is not uniform because the dimensions are too large, so the error increases. On the other hand, many methods are incapable of measuring the pore size of the pores, which are much smaller than the pores of nanofiber layers. SEM images are two-dimensional with layers stacked together. To consider pores the best method is nano-computed tomography (nano-CT) [36, 42, 43], but this device is not widely available. Its advantages include measuring pore size by machine and operator, providing completely accurate information about layer structure, and being very precise. Nanofiber imaging with micro-computed tomography (micro-CT) scanners is not possible due to the limitation of not being visible at distances less than 200 μm.

4. Conclusions

Methods for measuring and controlling porosity, as well as the importance of porosity in applications, are discussed in this chapter. The best method for measuring the porosity of the nanofiber layers was introduced as nano-CT. The focus of future work will be on controlling the porosity for various applications and making electrospun layers suitable for new applications. In addition, new porosity-controlling methods may be introduced in future works.

Conflict of interest

The authors declare that they have no competing interests.

Author details

Sedigheh Aghayari
Sharif University of Technology, Tehran, Iran

*Address all correspondence to: 1415he@gmail.com

IntechOpen

References

[1] Thenmozhi S, Dharmaraj N, Kadirvelu K, Kim HY. Electrospun nanofibers: New generation materials for advanced applications. Materials Science and Engineering B. 2017;**217**:36-48

[2] Anstey A, Chang E, Kim ES, Rizvi A, Kakroodi AR, Park CB, et al. Nanofibrillated polymer systems: Design, application, and current state of the art. Progress in Polymer Science. 2021;**113**:101346

[3] Ramakrishna S, Fujihara K, Teo WE, Yong T, Ma Z, Ramaseshan R. Electrospun nanofibers: Solving global issues. Materials Today. 2006;**9**(3):40-50

[4] Krifa M, Yuan W. Morphology and pore size distribution of electrospun and centrifugal forcespun nylon 6 nanofiber membranes. Textile Research Journal. 2016;**86**(12):1294-1306

[5] Yu Y, Ma R, Yan S, Fang J. Preparation of multi-layer nylon-6 nanofibrous membranes by electrospinning and hot pressing methods for dye filtration. RSC Advances. 2018;**8**(22):12173-12178

[6] Tan GZ, Zhou Y. Electrospinning of biomimetic fibrous scaffolds for tissue engineering: A review. International Journal of Polymeric Materials and Polymeric Biomaterials. 2020;**69**(15):947-960

[7] Ray SS, Chen SS, Nguyen NC, Nguyen HT. Electrospinning: A versatile fabrication technique for nanofibrous membranes for use in desalination. In: Nanoscale Materials in Water Purification; 2019. pp. 247-273

[8] Tornello PRC, Caracciolo PC, Cuadrado TR, Abraham GA. Structural characterization of electrospun micro/nanofibrous scaffolds by liquid extrusion porosimetry: A comparison with other techniques. Materials Science and Engineering: C. 2014;**41**:335-342

[9] Cipitria A, Skelton A, Dargaville TR, Dalton PD, Hutmacher DW. Design, fabrication and characterization of PCL electrospun scaffolds—a review. Journal of Materials Chemistry. 2011;**21**(26):9419-9453

[10] Xue J, Wu T, Dai Y, Xia Y. Electrospinning and electrospun nanofibers: Methods, materials, and applications. Chemical Reviews. 2019;**119**(8):5298-5415

[11] Frey MW, Li L. Electrospinning and porosity measurements of nylon-6/poly (ethylene oxide) blended nonwovens. Journal of Engineered Fibers and Fabrics. 2007;**2**(1):155

[12] Sreedhara SS, Tata NR. A novel method for measurement of porosity in nanofiber mat using pycnometer in filtration. Journal of Engineered Fibers and Fabrics. 2013;**8**(4):155

[13] Pham QP, Sharma U, Mikos AG. Electrospun poly (ε-caprolactone) microfiber and multilayer nanofiber/microfiber scaffolds: Characterization of scaffolds and measurement of cellular infiltration. Biomacromolecules. 2006;**7**(10):2796-2805

[14] Anand Ganesh V, Kundukad B, Cheng D, Radhakrishnan S, Ramakrishna S, Van Vliet KJ. Engineering silver-zwitterionic composite nanofiber membrane for bacterial fouling resistance. Journal of Applied Polymer Science. 2019;**136**(22):47580

[15] Hernandez C, Gupta SK, Zuniga JP, Vidal J, Galvan R, Martinez M, et al.

Performance evaluation of Ce3+ doped flexible PVDF fibers for efficient optical pressure sensors. Sensors and Actuators A: Physical. 2019;**298**:111595

[16] Ning J, Yang M, Yang H, Xu Z. Tailoring the morphologies of PVDF nanofibers by interfacial diffusion during coaxial electrospinning. Materials & Design. 2016;**109**:264-269

[17] Ramakrishna S, Fujihara K, Teo W. LTC. An introduction to eletrospinning and nanofibers. 2005

[18] Ho ST, Hutmacher DW. A comparison of micro CT with other techniques used in the characterization of scaffolds. Biomaterials. 2006;**27**(8):1362-1376

[19] Li L, Jiang Z, Li M, Li R, Fang T. Hierarchically structured PMMA fibers fabricated by electrospinning. RSC Advances. 2014;**4**(95):52973-52985

[20] Nayani K, Katepalli H, Sharma CS, Sharma A, Patil S, Venkataraghavan R. Electrospinning combined with nonsolvent-induced phase separation to fabricate highly porous and hollow submicrometer polymer fibers. Industrial & Engineering Chemistry Research. 2012;**51**(4):1761-1766

[21] Szewczyk PK, Gradys A, Kim SK, Persano L, Marzec M, Kryshtal A, et al. Enhanced piezoelectricity of electrospun polyvinylidene fluoride fibers for energy harvesting. ACS Applied Materials & Interfaces. 2020;**12**(11):13575-13583

[22] Casper CL, Stephens JS, Tassi NG, Chase DB, Rabolt JF. Controlling surface morphology of electrospun polystyrene fibers: Effect of humidity and molecular weight in the electrospinning process. Macromolecules. 2004;**37**(2):573-578

[23] Szentivanyi A, Chakradeo T, Zernetsch H, Glasmacher B. Electrospun cellular microenvironments: Understanding controlled release and scaffold structure. Advanced Drug Delivery Reviews. 2011;**63**(4-5):209-220

[24] Cao X, Chen W, Zhao P, Yang Y, Yu DG. Electrospun porous nanofibers: Pore–forming mechanisms and applications for photocatalytic degradation of organic pollutants in wastewater. Polymers. 2022;**14**(19):3990

[25] Joshi MK, Pant HR, Tiwari AP, Park CH, Kim CS. Multi-layered macroporous three-dimensional nanofibrous scaffold via a novel gas foaming technique. Chemical Engineering Journal. 2015;**275**:79-88

[26] Milleret V, Simona B, Neuenschwander P, Hall H. Tuning electrospinning parameters for production of 3D-fiber-fleeces with increased porosity for soft tissue engineering applications. European Cells & Materials. 2011;**21**(1473-2262):286-303

[27] Rosman N, Salleh WNW, Jamalludin MR, Adam MR, Ismail NH, Jaafar J, et al. Electrospinning parameters evaluation of PVDF-ZnO/Ag2CO3/Ag2O composite nanofiber affect on porosity by using response surface methodology. Materials Today: Proceedings. 2021;**46**:1824-1830

[28] Bandegi A, Moghbeli MR. Effect of solvent quality and humidity on the porous formation and oil absorbency of SAN electrospun nanofibers. Journal of Applied Polymer Science. 2018;**135**:45586

[29] Vitale A, Massaglia G, Chiodoni A, Bongiovanni R, Pirri CF, Quaglio M. Tuning porosity and functionality of electrospun rubber nanofiber mats by photo-crosslinking. ACS Applied Materials & Interfaces. 2019;**11**(27):24544-24551

[30] de Valence S, Tille JC, Giliberto JP, et al. Advantages of bilayered vascular grafts for surgical applicabilityand tissue regeneration. Acta Biomaterialia. 2012;**8**(11):3914-3920

[31] Sarna-Boś K, Skic K, Sobieszczański J, Boguta P, Chałas R. Contemporary approach to the porosity of dental materials and methods of its measurement. International Journal of Molecular Sciences. 2021;**22**(16):8903

[32] Ladjevardi SM, Asnaghi A, Izadkhast PS, Kashani AH. Applicability of graphite nanofluids in direct solar energy absorption. Solar Energy. 2013;**94**:327-334

[33] Øysæd H, Ruyter IE. Water sorption and filler characteristics of composites for use in posterior teeth. Journal of Dental Research. 1986;**65**(11):1315-1318

[34] Murray PE, García Godoy C, García Godoy F. How is the biocompatibilty of dental biomaterials evaluated? Medicina Oral, Patología Oral y Cirugía Bucal (Internet). 2007;**12**(3):258-266

[35] Guarino V, Causa F, Taddei P, Di Foggia M, Ciapetti G, Martini D, et al. Polylactic acid fibre-reinforced polycaprolactone scaffolds for bone tissue engineering. Biomaterials. 2008;**29**(27):3662-3670

[36] Su Z, Decencière E, Nguyen TT, El-Amiry K, De Andrade V, Franco AA, et al. Artificial neural network approach for multiphase segmentation of battery electrode nano-CT images. npj. Computational Materials. 2022;**8**(1):1-11

[37] Ghasemi-Mobarakeh L, Semnani D, Morshed M. A novel method for porosity measurement of various surface layers of nanofibers mat using image analysis for tissue engineering applications. Journal of Applied Polymer Science. 2007;**106**(4):2536-2542

[38] Karageorgiou V, Kaplan D. Porosity of 3D biomaterial scaffolds and osteogenesis. Biomaterials. 2005;**26**(27):5474-5491

[39] Kim DH, Bae J, Lee J, Ahn J, Hwang WT, Ko J, et al. Porous nanofiber membrane: Rational platform for highly sensitive thermochromic sensor. Advanced Functional Materials. 2022;**32**(24):2200

[40] Shi G, Cai Q, Wang C, Lu N, Wang S, Bei J. Fabrication and biocompatibility of cell scaffolds of poly (l-lactic acid) and poly (l-lactic-co-glycolic acid). Polymers for Advanced Technologies. 2002;**13**(3-4):227-232

[41] Topuz F, Abdulhamid MA, Holtzl T, Szekely G. Nanofiber engineering of microporous polyimides through electrospinning: Influence of electrospinning parameters and salt addition. Materials & Design. 2021;**198**:109280

[42] Lu X, Bertei A, Finegan DP, Tan C, Daemi SR, Weaving JS, et al. 3D microstructure design of lithium-ion battery electrodes assisted by X-ray nano-computed tomography and modelling. Nature Communications. 2020;**11**(1):1-13

[43] Aghayari S. A novel method for measuring the porosity of the nanowebs. Results in Materials. 2022;**2022**:100345

Section 3

Biotechnological Applications

Chapter 5

Antibacterial, Antifungal and Antiviral Nanocomposites: Recent Advances and Mechanisms of Action

Suchita C. Warangkar, Manish R. Deshpande, Narayan D. Totewad and Archana A. Singh

Abstract

Over the past ten years, there has been a significant increase in research into the study of nanocomposites. Nanocomposites vary in their physical and chemical properties. In today's era, eco-friendly, nontoxic, biocompatible, biobased fillers and composites should be synthesized to increase their societal value in various aspects. These materials have seen extensive use across several industries, from biosensors to biomedicine. Great strides have been made in the field of Microbiology, particularly as Antibacterial agents, among these applications. The objective of this review is to present a thorough analysis of several Nanocomposites that reveal promising antibacterial activity. Such Nanocomposites are reviewed in detail, as well as their antibacterial efficacy is discussed.

Keywords: antibacterial activities, antifungal activities, antiviral activities, nanocomposites, antibacterial agents

1. Introduction

A composite material is referred to as a nanocomposites when it contains a phase with nanoscale morphology, such as nanoparticles, nanotubes, or lamellar nanostructure. As a result of their numerous phases, they qualify as multiphase materials, and at least one of those phases must have a diameter between 10 and 100 nm. In order to get beyond the limits of present engineering materials, nanocomposites have developed as suitable substitutes. Nanocomposites may be categorized based on their dispersed matrix and dispersed phase materials [1]. Thanks to this rapidly emerging field, it is now possible to generate a wide range of exciting new materials with distinctive properties.

The interfacial and morphological features of the originals, together with their own characteristics, had a significant impact on the so-called found's characteristics. It is evident that, parent component parts are unaware of the newly created feature in the composite material and this intricate structure enhances its applicability [2, 3]. In order to create new materials with amazing flexibility and an increase in their physical properties, nanocomposites are based on the idea of employing building pieces that have dimensions in the nano scale range. Nanocomposites are made up of a bulk matrix and one or more nano dimensional phase(s) that differ from one another in terms of their chemical and structural makeup and properties. Inorganic nanoclusters, fullerenes, clays, and biological molecules can be mixed with a range of organic polymers, organic and organometallic chemicals, biological molecules, enzymes, and sol-gel produced polymers. Inorganic nanoclusters, fullerenes, clays, metals, oxides, or semiconductors can be mixed with a range of organic polymers, organic and organometallic chemicals, biological molecules, enzymes, and sol-gel produced polymers to produce nanocomposites.

2. Nanocomposites that reveal promising antibacterial activity

Antimicrobially active products are a recent development in nanoparticle-based materials that have gained significant attention. It has been documented that nanoscaled materials, such as fabrics, plastics, and metals coated with nano-silver, as well as nanocomponents based on titanium dioxide, magnesium oxide, copper, copper oxide, zinc oxide, cadmium selenide/telluride, chitosan, and carbon nanotubes, possess biocidal or bacteriostatic properties [4]. Both gram-positive and gram-negative bacteria, including *Escherichia coli* and *Pseudomonas aeruginosa*, have demonstrated the antibacterial activity of nanosized metal compounds. These bacteria include *Staphylococcus aureus* and *Bacillus subtilis*. The most frequently used antibacterial agents are nanomaterials with a silver base [5]. The antibacterial properties of metallic, ionic, and nanoscale silver compounds added to alumina nanopowder were described. However, gram-positive and gram-negative bacteria differ in their sensitivity to silver-doped nanocrystalline material [6].

Antimicrobially effective nanomaterials exist in the form of salts, oxides, complexes, and elemental nanoparticles. Because of their small size, chemical toxicity, and distinctive shape, they are effective at damaging cell membranes. Cell membrane surface loading and permeability may be disturbed as a result of nanoparticles' detrimental effects. Probably the most frequent method by which nanoparticles affect bacteria is through the production of reactive oxygen species (ROS) [7]. On "model" bacterial strains like *E. coli*, newly developed or modified nanoparticles' antimicrobial properties are typically tested. An assay for turbidity, a microdilution method, and the disc diffusion method are the main procedures. The antimicrobial activity of some industrial products was tested using a number of ISO regulations, such as ISO 20743 for textile products and ISO 22196 for plastics and other non-porous materials [8].

In tests with *E. coli*, *P. aeruginosa*, S. aureus, and *B. subtilis*, a potential wound healing nano-based material composed of genipin-crosslinked chitosan, poly (ethylene glycol), zinc oxide, and silver produced significant antibacterial activity [9]. With the addition of silver nanoparticles to hydrogels, researchers successfully achieved a significant antimicrobial impact for chitosan films, making them potentially useful in wound dressings as well [10]. The antimicrobial effect of the combination of silver nitrate and titanium dioxide nanoparticles applied to facemasks was described [11]. After 48 hours of testing, they noticed a 100% decrease in viable *E. coli* cells. The magnetic nanocomposite film created after dispersing magnetic nano—Fe_2O_3 in a chitosan matrix has potential uses in biosensors and tissue [12].

The use of highly toxic chemical reagents in the production of nanoparticles for medical applications should be avoided, especially when using materials containing silver (Ag). As a result, a "green synthesis approach" is taken into account. For instance, aqueous solutions of $AgNO_3$, glucose, and starch can be used to produce starch-protected Ag nanoparticles. Using these solutions, the reduction of Ag $(NH_3)^{2+}$ by carbohydrates results in the production of nano-Ag films (50–200 nm), Ag hydrosols (20–50 nm), and Ag colloids. Can also be used to reduce Ag $(NH_3)^{2+}$. In tests with *E. coli*, *P. aeruginosa, S. aureus*, and *B. subtilis*, a nano-based substance with the potential to treat wounds and containing genipin-crosslinked chitosan, poly (ethylene glycol), zinc oxide, and silver, demonstrated antibacterial activity [13].

Since chitosan nanoproducts have been shown to have potential antimicrobial properties, many medical applications of nanoproducts (chitosan-carbon dots, chitosan-i-cysteine quantum dots, chitosan-based biosensors and biomarkers) [14] are based on these materials. Chitosan-poly(N-vinylpyrrolidone)-TiO_2 Nanocomposite was proposed as a wound dressing material due to its significant antibacterial impact against *P. aeruginosa*, *E. coli*, S. aureus, and *B. subtilis*. In this study, the titanium dioxide nanocomponent was suggested to be responsible for the adsorption of bacteria and their inactivation [15].

By combining chitosan and 4-(ethoxycarbonyl) phenyl-1-amino-oxobutanoic acid with nano-Ag, Srivastava et al., 2011. created a nanocomposite. The effect of the obtained nano-film on bacteria like *S. aureus*, *E. coli*, and *P. aeruginosa* led to its proposal as a material for use in medicine [16]. Numerous medical applications benefit from the antibacterial properties of nanoparticle-based materials, including implants, wound and burn dressings, medical devices, filters, and dental plaque reduction materials. One of the most crucial justifications for the application of novel nanocomposites for clinical use is their potential impact on antibiotic-resistant bacteria, which pose a serious problem in current medical settings. In textile modification and impregnation, as well as "construction" elements for implants, cements, and resins, in the antibacterial coatings of external ventricular drains and venous catheters that lessen the risk of potential infections, nanocomposites can be used [4].

The most promising areas of nanotechnology applications are in the development of new antibacterial agents. Nanocluster engineering can broaden the application of Ag- and Au-based antimicrobial preparations. The commercial application of nanoproducts should also be carefully monitored because of their potential negative environmental effects. The use of Au, Ag, and Cu-based nanoclusters in medicine and biosensing is widespread. According to Zheng et al., 2016 the generation of ROS is the mechanism that most likely causes pathogenic bacteria to be destroyed in the presence of Ag-nanoclusters, whereas the core surface speciation of the nanoclusters may be related to their cellular toxicity. Due to the significantly higher surface-to-volume ratio they can achieve due to their ultra small size and interaction with intracellular components, they have a stronger antibacterial effect. It's possible that adding Ag nanoclusters to medications will enhance their therapeutic effects. In the presence of combined daptomycin-Ag nanoclusters, damage to microbial DNA was noted [17]. The nucleation and growth mechanism of Thiolate-protected Au nanoclusters with different topologies within the inner core of various clusters were described [18].

The development of polypyrrole-based nanocomposites as alternative antibacterial agents also represents a promising strategy to be applied against the prevailing multi-resistant bacteria. The composites are made up of different fillers (metal nanoparticles, carbon nanotubes, and polysaccharides) and strategies to improve their action (such as light and electrical stimulation) [19]. Graphene oxide–silver (Ag–GO) nanocomposite has emerged as a vital antibacterial agent very recently. It was successfully applied to *E. coli* to investigate antibacterial activity by varying its dose concentration. The functional groups of GO facilitated the binding of Ag nanoparticles to silver nanoparticles. The antibacterial properties of GO-Ag nanocomposite were studied using gram-negative *E. coli* ATCC 25922 and gram-positive *S. aureus* ATCC 6538 and showed excellent antibacterial activity. In this study, results demonstrated that GO-Ag nanocomposite, as a kind of antibacterial material, had great promise for application in a wide range of biomedical applications [20].

The Cu_2O-GO nanocomposites have rarely been studied before. The Cu_2O-GO nanocomposites show potent antibacterial activities against both *E. coli* and *S. aureus*. Bactericidal activity was also observed for the Cu based bionanocomposite samples against both gram-positive (*S. aureus*) and gram-negative (*Klebsiella pneumoniae*) bacteria. Enhancement of antibacterial activity was observed with increasing copper content in nanocomposites. Results confirm the potential of bionanocomposites containing copper nanostructures as new antimicrobial materials [21].

Cellulose/Ag nanocomposites were prepared using two distinct methodologies and two different cellulose substrates: vegetable and bacterial cellulose. Detailed studies on their antibacterial activity were conducted on *B. subtilis*, *S. aureus*, and *K. pneumoniae*. Silver nanoparticles present in these cellulosic fibers in concentrations as low as 5.0 wt% make them effective antibacterial materials [22].

The antibacterial activity of AuNPs-COOH/$AgNO_3$, $MnFe_2O_4$@SiO_2@Au and Bi_2S_3 nanocomposites against a wide range of gram-negative bacteria has been demonstrated. Green synthesis methods were applied and showed good activity against some gram-negative bacterial strains. A photocatalytic system comprising TNTs/Au/CDs was developed for a bactericidal approach. An Ag–Au/CeO_2 nanostructure was produced with a maximum zone of inhibition against *S. aureus* and *E. coli* strains. The produced nanohybrid showed acceptable antibacterial activity and was applicable for marine antifouling paint and sewage treatment. Au@TiO_2-NT was light-independent and applicable to the dark environment inside tissues, such as for orthopedic devices and implants [23, 24].

Nanocomposites and a composite based on poly (butylene adipate-co-terephthalate) were synthesized using commercial copper nanoparticles. The materials showed good inhibitory responses against the nonresistant strains *Enterococcus faecalis*, *Streptococcus mutans*, and *S. aureus*. They had the highest biocidal effect, even against resistant bacteria like *Acinetobacter baumannii* [25].

Magnetic cores loaded with metallic nanoparticles can be promising nano-carriers for successful drug delivery at infectious sites. The cobalt acetate was synthesized, and the decoration of AgNPs was carried out with silver acetate. The antibacterial performance of nanocomposites against *E. coli* and *B. subtilis* was found to be density-dependent. Silver nanocomposites exhibiting antiviral, antifungal, antibacterial, antiangiogenic, and antiinflammatory activities are discussed as potential candidates for several biomedical applications due to their ability to bind with the biomolecules of microbial cells [26].

In brief, the antibacterial GN/Ni $(OH)_2$ composite has been prepared using a facile method, providing powerful antibacterial capacity, good biocompatibility, and long-term effectiveness. The antibacterial activity of GN/Ni $(OH)_2$ was dose dependent and obviously exceeded that of rGO and GO. The GN/Ni$(OH)_2$ could efficiently kill gram-negative or positive bacteria with a low dose and exert low toxicity toward normal cells, motivating their potential safe applications. The results revealed that the improved 3D contact between GN/Ni$(OH)_2$ and bacteria enhanced the physical punctures on cells, causing severe leakage of intracellular components and leading to cell apoptosis. The GN/Ni$(OH)_2$ could not induce a remarkable increase in ROS

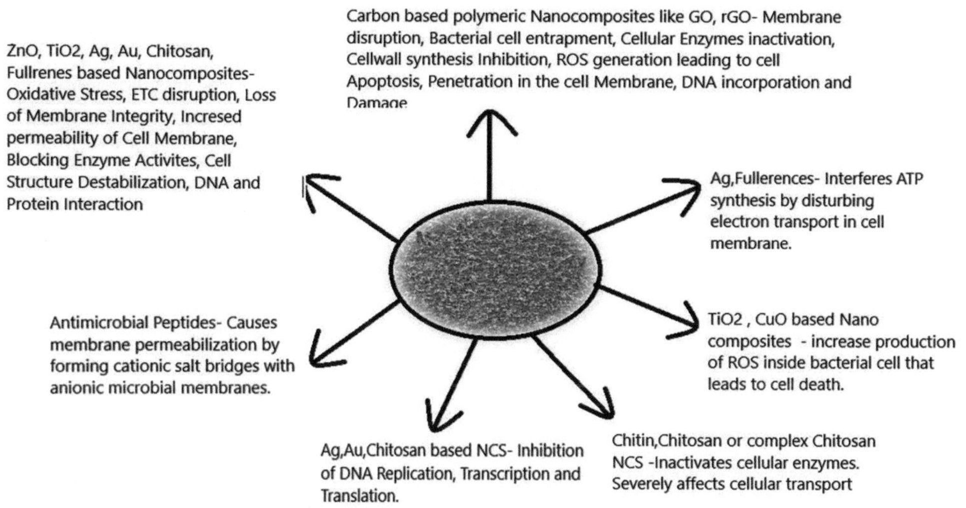

Figure 1.
Various modes of antibacterial and antifungal activities of different nanocomposites.

production, indicating that the ROS-dependent oxidative stress induced by GN/ $Ni(OH)_2$ will not affect antibacterial efficiency [27]. Nanocomposites that reveal various modes of antimicrobial Activities were enlisted in **Figure 1**.

3. Nanocomposites that reveal promising antifungal activity

Growing worldwide populations as a result of medical developments make more patients vulnerable to superficial and serious fungal infections. Dermatophytes like *Microsporum*, *Epidermophtyon*, and *Trichophyton*, as well as species from the genera *Candida*, *Aspergillosis*, and *Cryptococcus*, are among the fungi that are frequently linked to these disorders. Additionally, as the world's population rises, so do agricultural needs. Therefore, owing to food insecurity, fungal infections of preharvested crops and stored food by plant diseases such as *Magnaporthe oryzae* and *Fusarium oxysporum* might have negative socioeconomic impacts. The majority of current antifungal treatment plans are based on small molecule antifungal medications. Poor solubility and bioavailability, however, place restrictions on these medications. Additionally, there is an increase in antifungal resistance to these medications. A significant worldwide healthcare concern is the effect of fungal illnesses and the development of antimicrobial drugs against pathogenic fungus. Silver-loaded hydroxyapatite (Ag/HAP) nanocomposites (NCs) with varying Ag contents were tested against susceptible and resistant *Candida species* for their antifungal efficacy. *Candida krusei* had the inhibitory impact, followed by *Candida parapsilosis sensu stricto* and *Candida tropicalis* [28].

Antimicrobial polymers thus provide a different approach to combat fungi. The cationic regions of antifungal polymers have been reported to react with microbial phospholipids and membranes, and the hydrophobic areas are known to repel water. Such synthetic or natural polymers exhibit various antifungal activities, like fungal cell membrane permeabilization or fungal cell membrane depolarization. It might be challenging to determine their relative importance as Antifungal candidates. Due to these polymers' chemical structure, they can be coupled to provide synergistic effects with other antimicrobial substances such as metal ions, charcoal, lipids, and current antifungal medications. In certain instances, antifungal nanocomposites and polymers surpass typical small molecule antifungal drugs in terms of antifungal efficacy or toxicity [29, 30].

Cationic antimicrobial polymers and nanocomposites with antifungal activity as well as the state of knowledge on the antifungal mode of action were studied. The innate immune response includes antimicrobial peptides (AMPs), sometimes referred to as host defense peptides. The host is shielded against encroaching diseases by these substances, which are generated by plants, animals, and microbes. These peptides have short, amphiphilic sequences with an average length of 100 amino acids. The majority of AMPs are cationic, but those with high levels of histidine have powerful antifungal properties. Cathelicidins are an illustration of this. The primary storage location for this class of antifungal AMPs in macrophages is the lysosome, which is a component of the human innate immune system. However, certain AMPs with anionic charges need metal ions to be activated biologically. For membrane permeabilization, anionic AMPs bind metal ions to create cationic salt bridges with anionic microbial membranes. Even though certain anionic AMPs are credited with this mechanism, less is known about their antibacterial action than with cationic AMPs [31, 32].

In order to study the antifungal efficacy of these peptides against *Candida albicans*, Ramamourthy et al. (2020) synthesized peptides with various numbers of lysine and tryptophan repetitions (KWn-NH_2). The antifungal and biofilm-eradication abilities of these peptides increase with peptide length, with the longest peptide, KW5, exhibiting toxicity in a human keratinocyte cell line, while the smallest peptide, KW2, showed no antifungal activity by Ramamourthy et al., 2020. The membranes of fungus cells were not damaged by the KW4 peptide. However, KW4 was shown to be linked to fungal RNA in the cytoplasm of *C. albicans*, as revealed by laser-scanning Confocal Microscopy. This indicates that not all of these peptides antifungal mechanisms include membrane permeabilization. Instead, these synthetic AMPs localize within the cell where they disrupt cellular activities by attaching to specific receptors after entering fungal cells [33].

Despite synthetic AMPs showing broad-spectrum antifungal activity and low toxicity, research into antimicrobial polymers is often focused on synthetic polymers as they are considerably cheaper to produce in comparison and share functional cationic similarities with AMPs. Polyhexamethylene biguanide (PHMB) is a synthetic quarternary ammonium polymer which has been established to be an effective antimicrobial agent with the added advantages of low toxicity. It exhibits a high therapeutic index and broad-spectrum antifungal activity due to its biguanide groups and is commonly used as a preservative in cosmetics, water purification systems, and contact lens cleaning solutions. It is also used clinically for wound cleaning, where it shows excellent biocompatibility. Although PHMB shows membrane disruption abilities due to its phospholipid binding, the exact antifungal mechanism of action remains unclear. The antifungal mechanism is thought to involve cell wall destabilization and membrane permeabilization. Gene expression studies in *Saccharomyces cerevisiae* indicated an increase in the expression of cell wall integrity genes and protein kinase C, which regulates cell maintenance. This suggests PHMB also damages the β-glucan structure of the *S. cerevisiae* cell wall [34].

Polyethylenimines (PEI) are amine-containing polymers with a two-carbon (CH_2CH_2) spacer. At room temperature, they can be found in branched and linear polymeric forms in various states. They have been widely used in in vitro transfections and drug delivery. Such study has demonstrated depolarization of *C. albicans* membranes for its inactivation, though the precise antifungal mechanism is unknown [35, 36].

HTCC, a chitosan derivative used as a preservative in the cosmetics industry, is an antimicrobial polymer. In a study by Hoque et al. 2016, HTCC was shown to have strong antifungal activity (MIC = 125,250 g/ml), with killing occurring in just two hours. It targets the fungal cell membrane, increasing membrane permeability, similarly to chitosan, and exhibits extremely low toxicity (HC_{50} = > 100 g/ml) in a mouse model [37, 38].

Polymethacrylates (PMMA) is a polymer made of methacrylic acid esters. It is used as a wrinkle filler in cosmetics and as an intraocular lens in ophthalmology. Because it is used in medical devices, it is susceptible to microbial colonization by pathogens like *C. albicans*. Undecylenic acid (UA), a monounsaturated lipid with established antifungal activity at various UA concentrations (3–12%), changed the surface of PMMA by making it more hydrophilic. At UA concentrations of 6%, *C. albicans* exposed to these PMMA-UA composites exhibited decreased attachment, growth, and increased death of these fungal cells. Despite the fact that UA concentrations of 9% showed a 95% eradication of *C. albicans*, these composites were extremely toxic to human cells, with a 50% reduction in viability [39].

Antimicrobial metal nanoparticles (NPs) are used more frequently than antimicrobial drugs because they reduce the risk of antimicrobial resistance. They do, however, have a toxic reputation. They can be incorporated into biomaterials like proteins, peptides, and sugars to increase biocompatibility. The Ag-Au alloy nanoparticles (Ag-AuNPs) with potential broad-spectrum uses, such as a coating material for medical devices or for drug delivery, were created using a green synthesis approach [7].

The concentration of Ag used in the synthesis of NPs determines its size and shape. Due to the silver ions, these nanocomposites effectively combatted *C. albicans*. The Ag-Au alloy nanoparticles, however, displayed improved performance. Due to its broad-spectrum antibacterial activity and low toxicity toward mammalian cells, cotton is the most commonly used natural fiber for textiles today. Silver nanoparticles have recently been added to it [40].

Likewise functionalized fabric treated with a nanocomposite made of silver nanoparticles and carboxymethyl chitosan had excellent antifungal and antibacterial properties against *C. albicans* and *Aspergillus niger* (AgNPs-CMC). This fabric was functionalized to demonstrate how this fabric could be used to make hospital clothing to lower nosocomial infections. In this research, reported synthesis of silver-incorporated Chitosan nanocomposites (Ag@CS), CS was used as a reducing and stabilizing agent. The fungicide Antracol (An) was then combined with Ag@CS/An to effectively combat *Phytophthora capsici*. Researchers discovered that Ag@CS/An was found to have significantly stronger and synergistic antifungal ability than Ag@CS nanocomposites or Antracol nanocomposites, which had diameters upto 44.6 nm [41]. The TiO_2-NPs, in particular, are also effective as antimicrobial agents due to their high aspect ratio, large surface-to-volume ratio, and reactivity [42].

Nanomaterials made upof metals or metal oxides are produced using living things or their components. AgNPs against *Candida glabrata* were made in spherical or rod shapes with crystal structures made of 80% anatase and 20% rutile. Under spherical AgNPs (with a diameter range of 1–24 nm) produced by a filtrated suspension of *Aspergillus sydowii* fungi, *C. glabrata* has demonstrated a minimum inhibitory concentration of 0.125 ppm [43]. Additionally, a hydrothermal technique was used to decorate TiO_2@ZnO nanocomposites with AuNPs that shown antifungal activity against *C. albicans* (MTCC 282) and an antiproteinase activity [44].

Hesperidin, a flavanone disaccharide extracted from orange peel, was used to make ZnONPs, which demonstrated notable antiviral activity against the *Hepatitis A* virus and *Respiratory Syndrome CoronaVirus* 2 (SARS-CoV-2). They also displayed activity that was suitable for treating HIV infection (50% inhibition at 100 ppm) [45].

Seven novel silver chromite nanocomposites were synthesized and assayed to evaluate their antimicrobial, antiviral, and cytotoxic activities. Five bacterial species were used in this study: three gram-positive (*B. subtilis*, *Micrococcus luteus*, and *S. aureus*) and two gram-negative (*E. coli*, and *Salmonella enterica*). Three fungal species were also tested: *C. albicans*, *A. niger*, and *Aspergillus flavus*. The MIC of the tested compounds was determined using the bifold serial dilution method. These tested compounds could be attractive and alternative antibacterial compounds that open a new path in chemotherapy [46].

A sustainable and green method was used to prepare silver nanoparticles (Ag-NPs), followed by their incorporation into a tertiary nanocomposite consisting of starch, oxidized cellulose, and ethyl cellulose. Ag-NC significantly suppressed the growth of tested bacterial strains (*E. coli, P. aeruginosa, S. aureus,* and *B. subtilis*) as compared with controls. It has also exhibited antiviral effects against *Herpes Simplex Virus*, *Adenovirus* and *Coxsackie B Virus* in a dose-dependent manner. In conclusion, the prepared tertiary Ag-NCs had promising antibacterial, antifungal, as well as antiviral activities [7, 40].

Binary TiO_2/AgBr nanocomposites were synthesized using a facile ultrasonic irradiation route and characterized by various instruments. After adding AgBr nanoparticles, the antifungal activity was markedly enhanced. Silver ions in AgBr have a broad antimicrobial spectrum and can inhibit the growth of fungi. A sample with 20% of silver bromide represented the highest inhibitory concentration for the mycelial growth of *F. graminearum* and *S. sclerotiorum*. The inactivation rate decreased with increasing ultrasound irradiation time [47].

The negative effects of various biotic and/or biotic stresses on plants may be mitigated by silicon and its nanomaterials. For regulating the growth parameters and yield of *faba beans* infected with *Botrytis cinerea*, the antifungal role of silver/silicon dioxide nanocomposite (Ag/SiO_2NC) biosynthesized using a free-cell supernatant of *E. coli* was examined. In vitro tests revealed significant in vitro activity with a minimal inhibitory concentration (MIC) of 40 ppm. These are all encouraging findings for the use of the biosynthesized Ag/SiO_2NC as a secure and efficient antifungal agent against *B. cinerea* [48].

Nanocomposites that reveal various modes of antimicrobial inhibition enlisted in **Figure 1**.

4. Nanocomposites that reveal promising antiviral activity

The most recent research on viruses-designed coating materials as well as potential nanocoatings to stop the spread of the contagious SARS-COV-2 virus in response to the global health outbreak was well explored. Due to viral adhesion/colonization, subsequent proliferation, and biofilm formation, the exposed surfaces are contaminated. Hence, surface contamination possibly removed using the traditional disinfecting cleaning method, but studies shown that disinfecting only offers a momentary relief. The field of antiviral coating has seen some promising work, but more study is undoubtedly needed. In the creation of antiviral coatings, it is thought that nanomaterials like metal oxide nanostructures, Graphene, Carbon Nano Tube, Carbon quantum dots, Titanium dioxide, bio-based nanoparticles like chitosan, capped silver, Graphene, Gold and Silicon nanoparticles could play a key role [49, 50].

A team of researchers has shown that surface-adsorbed viruses can be effectively removed from surfaces using nanoparticles. The antiviral effect, which most likely results from a "contact killing mechanism," is highly dependent on the type of polymer and the affinity of the nanoparticles for the polymer. In this regard surface coatings made of nanocomposite materials with a polymer matrix and Cu/CuO nanoparticles synthesized and shown that surface-adsorbed viruses could be effectively removed [51].

Numerous viruses can survive and maintain their infectiousness on plastics for several days when exposed to ambient conditions. Measuring the persistence of various virus types on frequently used composite materials, like carbon-epoxy and glass-polyester laminates, is necessary. The polymer composites community has received a clear message from the SARS-CoV-2 global pandemic that there are opportunities for next-generation materials with virus-resistant surfaces [52].

Polymer Nano-Composites (PNC), such as polysaccharides nanocomposites, may play a significant role in the development of an antiviral drug for Covid-19. PNC could manage the health system, reduce lockdown times, and reduce social isolation while also saving money and energy [53].

The unique structure of graphene oxide sheets could contribute to the inhibition of infection by feline coronavirus with a lipid envelope. GO sheets with silver particles exhibited antiviral activity against both enveloped viruses and non-enveloped viruses. Negatively charged GO can absorb positively charged lipid membranes and induce rupture of membranes. The interactions between GO and the membrane can attract the absorption of more lipid membranes [54].

This study, proposed a blocking strategy against model respiratory viruses, severe acute respiratory syndrome coronavirus 2 (SARS-CoV-2) pseudovirus and porcine reproductive and respiratory syndrome virus (PRRSV) (PRRSV and SARS-CoV-2) infection by heparan sulfate analogue-modified two-dimensional (2D) transition metal carbides (MXenes) nanocomposites. The functional 2D nanocomposites with excellent physicochemical properties and abundant heparin analogue (MPS) demonstrated several unique advantages for antiviral research. Firstly, the Ti_3C_2-Au-MPS nanocomposites with a relatively uniform particle size and excellent biocompatibility can be synthesized in a facile method. Secondly, Ti_3C_2-Au-MPS nanocomposites can block PRRSV infection by inactivating PRRS virions in vitro and inhibiting its adsorption and invasion in host cells. Thirdly, Ti3C2-Au MPS nanocomposites have a potent inhibitory effect on SARS-CoV-2 infection, suggesting that these materials have broadspectrum antiviral activity against PRRSV and SARS-CoV-2 [55].

According to a study, aqueous medium was used to create Ag NP/Chitosan composites with antiviral activity against the Influenza A virus. Unreacted Ag NPs were not found in the composites, which were obtained as yellow or brown flocs. The experimental results demonstrate that virions and composites interacted. The synthesis methods control the antiviral and cytotoxic properties of the silver nanoparticle or nanocomposite by modifying its size, shape, morphology, and surface charge. As discussed in this work, biological approaches have emerged as a result of the shortcomings of physical and chemical approaches [56].

Antibacterial, antifungal, and antiviral properties of ZnO NPs and Activated Carbon nanoparticles were synthesized. On the human WI38 cell line, their

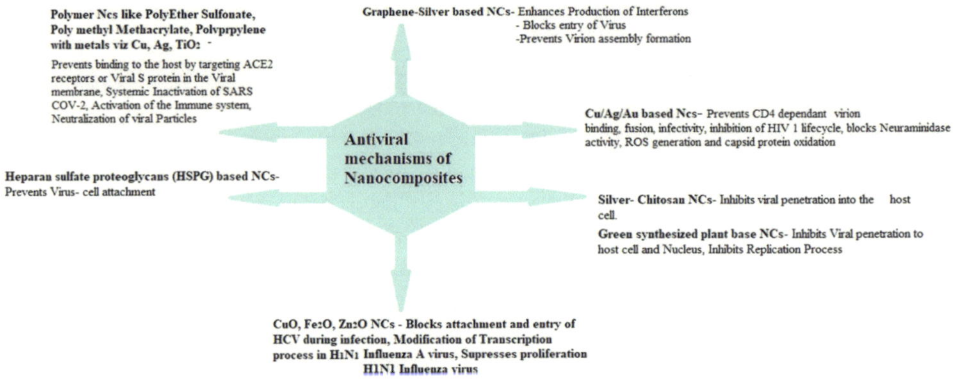

Figure 2.
Various modes of antiviral activities of different nanocomposites.

cytotoxicity was tested. Such nanocomposites reported as lethal at minimally toxic concentration reduced the Herpes Simplex Virus1 count by about 83%. (MNTC) [57].

The suggested TiO2 PL-DNA nanocomposites can be used to effectively and specifically inhibit different subtypes of influenza A virus. The proposed TiO_2 PL-DNA nanocomposites have remarkable antiviral activity, making them excellent platforms for drug development against a wide range of nucleic acid-related diseases, from infectious diseases to hereditary disorders [58]. Various Nanocomposites discussed above have shown antiviral activities against respected viruses. Possible Mechanisms of Antiviral activities were enlisted in **Figure 2**.

5. Conclusion

Antibacterial nanocomposites incorporating inorganic nanoparticles present higher antibacterial activity compared with their bulk counterparts due to their higher surface-to-volume ratio, resulting in improved contact with microorganisms. Antibacterial properties have been usually tested on nonpathogenic bacterial strains like *E. coli* and *S. aureus* as model organisms, but research should focus on other bacterial pathogens of different families. This would account for the increasing antibiotic resistance among various bacteria and their association as a severe hazard to worldwide public health. Some Graphene like nanocomposites with small-sized NPs are more effective against Gram-negative bacteria since they have larger surface area in contact with the bacteria. Well-dispersed nanomaterials show stronger antibacterial activity than the aggregated ones. The main challenge is obtaining reliable information on the interaction between bacteria and nano-structures. Another challenge is to analyze the toxicity associated with them. Antifungal polymers can be combined with other anti-microbial compounds to enhance their antibacterial activity. This flexibility provides great promise for applications that range from postharvest food preservation to healthcare, according to the World Health Organization (WHO). The potential of antifungal composites to replace antifungal drugs still remained unexplored. Likewise viral infections are difficult to treat because viruses spread and multiply quickly. Numerous new, deadly viruses, including the *Coronavirus, Ebola virus, Dengue virus, HIV,* and *Influenza virus*, are already causing chaos on people. Moreover Silver and biobased polymers as Nanocomposites which are selective antimicrobial compounds should be explored in future to enhance its applicability. Silver biobased nanocomposites are thought to be a powerful and cutting-edge pharmacological agent with strong antiviral activity against these viruses and microorganisms. This is to be thoroughly studied in all respects.

Acknowledgements

The Authors would like to thank Head, Department of Microbiology and Head, Department of Physics, Principal of our college and University stake holders for their cooperation.

Conflict of interest

The authors declare that there is no conflict of Interest regarding this publication.

Author details

Suchita C. Warangkar[1]*, Manish R. Deshpande[2], Narayan D. Totewad[3] and Archana A. Singh[4]

1 Department of Microbiology, Netaji Subhashchandra Bose Arts, Commerce and Science College, Swami Ramanand Teerth Marathwada University, Nanded, MS, India

2 Department of Physics, Netaji Subhashchandra Bose Arts, Commerce and Science College, Swami Ramanand Teerth Marathwada University, Nanded, MS, India

3 Department of Microbiology, B. K. Birla College of Arts, Science & Commerce (Autonomous), Kalyan, MS, India

4 Department of Chemistry, B. K. Birla College of Arts, Science & Commerce (Autonomous), Kalyan, MS, India

*Address all correspondence to: suchitawarangkar@gmail.com

References

[1] Billah R. Composites and nanocomposites. In: Jafar Mazumder M, Sheardown H, Al-Ahmed A, editors. Functional Polymers. Polymers and Polymeric Composites: A Reference Series. Cham: Springer; 2019. DOI: 10.1007/978-3-319-92067-2_15-1

[2] Camargo PHC, Satyanarayana KG, Wypych F. Nanocomposites: Synthesis, structure, properties and new application opportunities. Materials Research. 2009;**12**(1):1-39. DOI: 10.1590/s1516-14392009000100002

[3] Sen M. Nanocomposite Materials. In: Sen M, editor. Nanotechnology and the Environment. London, England: IntechOpen; 2020 https://www.intechopen.com/chapters/72636doi

[4] Agnihotri S, Dhiman NK. Development of Nano-antimicrobial biomaterials for biomedical applications. In: Advanced Structured Materials. Singapore: Springer Singapore; 2017. pp. 479-545 https://www.ncbi.nlm.nih.gov/pmc/articles/PMC7122509/

[5] Wang L, Hu C, Shao L. The antimicrobial activity of nanoparticles: Present situation and prospects for the future. International Journal of Nanomedicine. 2017;**12**:1227-1249. DOI: 10.2147/IJN.S121956

[6] Sadiq IM, Chowdhury B, Chandrasekaran N, Mukherjee A. Antimicrobial sensitivity of Escherichia coli to alumina nanoparticles. Nanomedicine. 2009;**5**(3):282-286. DOI: 10.1016/j.nano.2009.01.002

[7] Sánchez-López E, Gomes D, Esteruelas G, Bonilla L, Lopez-Machado AL, Galindo R, et al. Metal-based nanoparticles as antimicrobial agents: An overview. Nanomaterials (Basel). 2020;**10**(2):292. DOI: 10.3390/nano10020292

[8] Farouk MM, El-Molla A, Salib FA, Soliman YA, Shaalan M. The role of silver nanoparticles in a treatment approach for multidrug-resistant salmonella species isolates. International Journal of Nanomedicine. 2020;**15**:6993-7011. DOI: 10.2147/IJN.S270204

[9] Li S, Dong S, Xu W, Tu S, Yan L, Zhao C, et al. Antibacterial hydrogels. Advanced Science (Weinh). 2018;**5**(5):1700527. DOI: 10.1002/advs.201700527

[10] Wang Y, Dou C, He G, Ban L, Huang L, Li Z, et al. Biomedical potential of ultrafine Ag nanoparticles coated on poly (gamma-glutamic acid) hydrogel with special reference to wound healing. Nanomaterials (Basel). 2018;**8**(5):324, 1-11. DOI: 10.3390/nano8050324

[11] Li Y, Leung P, Yao L, Song QW, Newton E. Antimicrobial effect of surgical masks coated with nanoparticles. The Journal of Hospital Infection. 2006;**62**(1):58-63 https://www.sciencedirect.com/science/article/pii/S0195670105002069

[12] Hoque MA, Ahmed MR, Rahman GT, Rahman MT, Islam MA, Khan MA, et al. Fabrication and comparative study of magnetic Fe and α-Fe2O3 nanoparticles dispersed hybrid polymer (PVA + chitosan) novel nanocomposite film. Results in Physics. 2018;**10**:434-443 https://www.sciencedirect.com/science/article/pii/S2211379718309963

[13] Iravani S, Korbekandi H, Mirmohammadi SV, Zolfaghari B. Synthesis of silver nanoparticles:

Chemical, physical and biological methods. Research in Pharmaceutical Sciences. 2014;**9**(6):385-406 https://www.ncbi.nlm.nih.gov/pmc/articles/PMC4326978/

[14] Abourehab MAS, Pramanik S, Abdelgawad MA, Abualsoud BM, Kadi A, Ansari MJ, et al. Recent advances of chitosan formulations in biomedical applications. International Journal of Molecular Sciences. 2022;**23**(18):10975. DOI: 10.3390/ijms231810975

[15] Karwowska E. Antibacterial potential of nanocomposite-based materials – A short review. Nanotechnology Reviews. 2017;**6**(2):243-254. DOI: 10.1515/ntrev-2016-0046

[16] Srivastava R, Tiwari DK, Dutta PK. 4-(Ethoxycarbonyl) phenyl-1-amino-oxobutanoic acid-chitosan complex as a new matrix for silver nanocomposite film: Preparation, characterization and antibacterial activity. International Journal of Biological Macromolecules. 2011;**49**(5):863-870. DOI: 10.1016/j.ijbiomac.2011.07.015

[17] Zheng K, Setyawati MI, Lim T-P, Leong DT, Xie J. Antimicrobial cluster bombs: Silver nanoclusters packed with daptomycin. ACS Nano. 2016;**10**(8):7934-7942. DOI: 10.1021/acsnano.6b03862

[18] Liu C, Pei Y, Sun H, Ma J. The nucleation and growth mechanism of thiolate-protected Au nanoclusters. Journal of the American Chemical Society. 2015;**137**(50):15809-15816. DOI: 10.1021/jacs.5b09466

[19] da Silva Júnior FAG, Vieira SA, de Botton SA, da Costa MM, de Oliveira HP. Antibacterial activity of polypyrrole-based nanocomposites: A mini-review. Polímeros. 2020;**30**(4):e2020048, 1-9. DOI: 10.1590/0104-1428.08020

[20] Tang J, Chen Q, Xu L, Zhang S, Feng L, Cheng L, et al. Graphene oxide–silver nanocomposite as a highly effective antibacterial agent with species-specific mechanisms. ACS Applied Materials & Interfaces. 2013;**5**(9):3867-3874. DOI: 10.1021/am4005495

[21] Li M, Chen Z, Yang L, Li J, Xu J, Chen C, et al. Antibacterial activity and mechanism of GO/Cu2O/ZnO coating on ultrafine glass fiber. Nanomaterials (Basel). 2022;**12**(11):1857. DOI: 10.3390/nano12111857

[22] Pinto RJB, Marques PAAP, Neto CP, Trindade T, Daina S, Sadocco P. Antibacterial activity of nanocomposites of silver and bacterial or vegetable cellulosic fibers. Acta Biomaterialia. 2009;**5**(6):2279-2289 https://www.sciencedirect.com/science/article/pii/S1742706109000646

[23] Ahmad MA, Aslam S, Mustafa F, Arshad U. Synergistic antibacterial activity of surfactant free Ag–GO nanocomposites. Scientific Reports. 2021;**11**(1):1-9. DOI: 10.1038/s41598-020-80013-w

[24] Mobed A, Hasanzadeh M, Seidi F. Anti-bacterial activity of gold nanocomposites as a new nanomaterial weapon to combat photogenic agents: Recent advances and challenges. RSC Advances. 2021;**11**(55):34688-34698. DOI: 10.1039/d1ra06030a

[25] Jaramillo AF, Riquelme SA, Sánchez-Sanhueza G, Medina C, Solís-Pomar F, Rojas D, et al. Comparative study of the antimicrobial effect of nanocomposites and composite based on poly(butylene adipate-co-terephthalate) using Cu and Cu/Cu2O nanoparticles and CuSO4. Nanoscale Research Letters. 2019;**14**(1):158. DOI: 10.1186/s11671-019-2987-x

[26] Kanwal Z, Raza MA, Riaz S, Manzoor S, Tayyeb A, Sajid I, et al.

Synthesis and characterization of silver nanoparticle-decorated cobalt nanocomposites (Co@AgNPs) and their density-dependent antibacterial activity. Royal Society Open Science. 2019;**6**(5):182135. DOI: 10.1098/rsos.182135

[27] Li N, Zeng C, Qin Q, Zhang B, Chen L, Luo Z. Powerful antibacterial activity of graphene/nanoflower-like nickelous hydroxide nanocomposites. Nanomedicine (London, England). 2018;**13**(22):2901-2916. DOI: 10.2217/nnm-2018-0200

[28] Gottardo B, Lemes TH, Byzynski G, Paziani MH, von-Zeska-Kress MR, de Almeida MTG, et al. One-pot synthesis and antifungal activity of nontoxic silver-loaded hydroxyapatite nanocomposites against candida species. ACS Applied Nano Materials. 2019;**2**(4):2112-2120. DOI: 10.1021/acsanm.9b00091

[29] Ntow-Boahene W, Cook D, Good L. Antifungal polymeric materials and nanocomposites. Frontiers in Bioengineering and Biotechnology. 2021;**9**:780328. DOI: 10.3389/fbioe.2021.780328

[30] Velazco-Medel MA, Camacho-Cruz LA, Lugo-González JC, Bucio E. Antifungal polymers for medical applications. Medical Devices & Sensors. 2021;**4**(1):e10134, 1-23. DOI: 10.1002/mds3.10134

[31] Rank LA, Walsh NM, Liu R, Lim FY, Bok JW, Huang M, et al. A cationic polymer that shows high antifungal activity against diverse human pathogens. Antimicrobial Agents and Chemotherapy. 2017;**61**(10):e00204-17. DOI: 10.1128/AAC.00204-17

[32] Phoenix DA, Dennison SR, Harris F. Anionic Antimicrobial Peptides. In: Antimicrobial Peptides. Weinheim, Germany: Wiley-VCH Verlag GmbH & Co. KGaA; 2013. pp. 83-113. DOI: 10.1002/9783527652853.ch3

[33] Ramamourthy G, Park J, Seo C, Vogel J, H, Park Y. Antifungal and antibiofilm activities and the mechanism of action of repeating lysine-tryptophan peptides against *Candida albicans*. Microorganisms. 2020;**8**(5):758. DOI: 10.3390/microorganisms8050758

[34] Worsley A, Vassileva K, Tsui J, Song W, Good L. Polyhexamethylene biguanide: Polyurethane blend nanofibrous membranes for wound infection control. Polymers (Basel). 2019;**11**(5):915. DOI: 10.3390/polym11050915

[35] Zakeri A, Kouhbanani MAJ, Beheshtkhoo N, Beigi V, Mousavi SM, Hashemi SAR, et al. Polyethylenimine-based nanocarriers in co-delivery of drug and gene: A developing horizon. Nano Reviews & Experiments. 2018;**9**(1):1488497. DOI: 10.1080/20022727.2018.1488497

[36] Elnashar M. Immobilized molecules using biomaterials and nanobiotechnology. Journal of Biomaterials and Nanobiotechnology. 2010;**1**(01):61-77. DOI: 10.4236/jbnb.2010.11008

[37] Cele ZED, Somboro AM, Amoako DG, Ndlandla LF, Balogun MO. Fluorinated quaternary chitosan derivatives: Synthesis, characterization, antibacterial activity, and killing kinetics. ACS Omega. 2020;**5**(46):29657-29666. DOI: 10.1021/acsomega.0c01355

[38] Elnashar MM, Abdel L, Awad HM, Mouneir SM. Chitosan-benzofuran adduct for potential biomedical applications: Improved antibacterial and antifungal properties. Der Pharmacia Letter. 2015;7:107-117

[39] Petrović M, Bonvin D, Hofmann H, Mionić EM. Fungicidal PMMA-Undecylenic Acid Composites. International Journal of Molecular Sciences. 2018;**19**(1):184. DOI: 10.3390/ijms19010184

[40] Arenas-Chávez CA, de Hollanda LM, Arce-Esquivel AA, Alvarez-Risco A, Del-Aguila-Arcentales S, Yáñez JA, et al. Antibacterial and antifungal activity of functionalized cotton fabric with nanocomposite based on silver nanoparticles and carboxymethyl chitosan. Processes (Basel). 2022;**10**(6):1088. DOI: 10.3390/pr10061088

[41] Le VT, Bach LG, Pham TT, Le NTT, Ngoc UTP, Tran D-HN, et al. Synthesis and antifungal activity of chitosan-silver nanocomposite synergize fungicide against Phytophthora capsici. Journal of Macromolecular Science, Part A Pure and Applied Chemistry. 2019;**56**(6):522-528. DOI: 10.1080/10601325.2019.1586439

[42] Díez-Pascual A. Recent progress in antimicrobial nanomaterials. Nanomaterials (Basel) [Internet]. 2020;**10**(11):2315. DOI: 10.3390/nano10112315

[43] López de Dicastillo C, Guerrero Correa M, Martínez FB, Streitt C, José Galotto M. Antimicrobial effect of titanium dioxide nanoparticles. In: Mareș M, Lim SHE, Lai K-S, Cristina R-T, editors. Antimicrobial Resistance [Working Title]. London, England: IntechOpen; 2020. https://www.intechopen.com/chapters/70919. DOI: 10.5772/intechopen.90891

[44] Amiri MR, Alavi M, Taran M, Kahrizi D. Antibacterial, antifungal, antiviral, and photocatalytic activities of TiO2 nanoparticles, nanocomposites, and bio-nanocomposites: Recent advances and challenges. Journal of Public Health Research. 2022;**11**(2):227990362211041. DOI: 10.1177/22799036221104151

[45] Attia GH, Moemen YS, Youns M, Ibrahim AM, Abdou R, El Raey MA. Antiviral zinc oxide nanoparticles mediated by hesperidin and in silico comparison study between antiviral phenolics as anti-SARS-CoV-2. Colloids and Surfaces. B, Biointerfaces. 2021;**203**:111724. DOI: 10.1016/j.colsurfb.2021.111724 Epub 2021 Mar 26

[46] Sayed MA, El-Rahman TMAA, Abdelsalam HK, Ali AM, Hamdy MM, Badr YA, et al. Attractive study of the antimicrobial, antiviral, and cytotoxic activity of novel synthesized silver chromite nanocomposites. BMC Chemistry. 2022;**16**(1):39. DOI: 10.1186/s13065-022-00832-y

[47] Habibi-Yangjeh A, Davari M, Manafi-Yeldagermani R, Alikhah Asl S, Enaiati S, Ebadollahi A, et al. Antifungal activity of TiO2/AgBr nanocomposites on some phytopathogenic fungi. Food Science & Nutrition. 2021;**9**(7):3815-3823. DOI: 10.1002/fsn3.2357

[48] Baka ZA, El-Zahed MM. Antifungal activity of silver/silicon dioxide nanocomposite on the response of faba bean plants (Vicia faba L.) infected by Botrytis cinerea. Bioresources and Bioprocessing. 2022;**9**(1):102, 1-19. DOI: 10.1186/s40643-022-00591-7

[49] Toledo E, Dim S, Edri A, Greenshpan Y, Ottolenghi A, Eisner N, et al. Nanocomposite coatings for the prevention of surface contamination by coronavirus. PLoS One. 2022;**17**(8):e0272307. DOI: 10.1371/journal.pone.0272307

[50] Dahanayake MH, Athukorala SS, Jayasundera ACA. Recent breakthroughs in nanostructured antiviral coating and filtration materials: A brief review. RSC Advances. 2022;**12**(26):16369-16385

[51] Mouritz AP, Galos J, Linklater DP, Ladani RB, Kandare E, Crawford RJ, et al. Towards antiviral polymer composites to combat COVID-19 transmission. Nano Select. 2021;**2**(11):2061-2071. DOI: 10.1002/nano.202100078

[52] Akram JJ. Possible role of biomedical polymers in COVID-19 journey: A short review. Journal of Infectious Diseases and Epidemiology. 2021;7(4):201. DOI: 10.23937/2474-3658/1510201

[53] Tong T, Tang W, Xiao S, Liang J. Antiviral effects of heparan sulfate analogue-modified two-dimensional MXene nanocomposites on PRRSV and SARS-CoV-2. Advanced Nanobiomed Research. 2022;**2**(10):2200067. DOI: 10.1002/anbr.202200067

[54] Chen YN, Hsueh YH, Hsieh CT, Tzou DY, Chang PL. Antiviral activity of graphene-silver nanocomposites against non-enveloped and enveloped viruses. International Journal of Environmental Research and Public Health. 2016;**13**(4):430. DOI: 10.3390/ijerph13040430

[55] Mori Y, Ono T, Miyahira Y, Nguyen VQ, Matsui T, Ishihara M. Antiviral activity of silver nanoparticle/chitosan composites against H1N1 influenza a virus. Nanoscale Research Letters. 2013;**8**(1):93. DOI: 10.1186/1556-276X-8-93

[56] Jeevanandam J, Krishnan S, Hii YS, Pan S, Chan YS, Acquah C, et al. Synthesis approach-dependent antiviral properties of silver nanoparticles and nanocomposites. Journal of Nanostructure in Chemistry. 2022;**12**(5):809-831. DOI: 10.1007/s40097-021-00465-y

[57] Hassan HS, Abol-Fotouh D, Salama E, Elkady MF. Assessment of antimicrobial, cytotoxicity, and antiviral impact of a green zinc oxide/activated carbon nanocomposite. Scientific Reports. 2022;**12**(1):8774. DOI: 10.1038/s41598-022-12648-w

[58] Levina AS, Repkova MN, Bessudnova EV, Filippova EI, Mazurkova NA, Zarytova VF. High antiviral effect of $TiO2·PL$-DNA nanocomposites targeted to conservative regions of (−)RNA and (+)RNA of influenza a virus in cell culture. Beilstein Journal of Nanotechnology. 2016;**7**: 1166-1173. DOI: 10.3762/bjnano.7.108

Chapter 6

Perspective Chapter: Tissue-Electronics Interfaces

Shahab Ahmadi Seyedkhani and Raheleh Mohammadpour

Abstract

Tissue-electronics interfaces provide a two-way communication between biological tissue and external electronics devices to record electrophysiological signals and stimulation of the living organs. This chapter presents an overview of significant progresses in tissue-electronics interfaces. At first, we evaluate principal properties of the living tissue microenvironment important for tissue-specific equipment design. Next, we study charge transfer mechanisms in the biological tissues, bulk electrode materials, and tissue-electronics interfaces. After that, we highlight the current developing and promising advanced biomaterials for the neural electrodes, significantly leading to the development of bionanoelectronics and bionic organs. Finally, the challenges and future outlook of the neural interfaces will be discussed.

Keywords: bioelectronics, neural interfaces, biomaterials, composites, neural recording, electrical stimulation, charge transfer

1. Introduction

The discovery that human cells are capable of producing electrical signals and responding to electrical stimulation, has encouraged researchers to develop technologies based on monitoring the body's electrophysiological activities and electrical stimulation of living tissues. To understand that how the electronic systems interact efficiently with biological tissues, dominating over the structural complexity and function of the host tissue is essential. Biological tissue often contains the cells distributed in an extracellular matrix (ECM). The ECM is a specific biochemical composition consisting of various sugars and proteins in an aqueous medium. In addition to mechanical support, the ECM has biochemical and topological properties that affect the cellular functions such as migration, proliferation, differentiation and growth mechanisms. The properties of the ECM are different according to the type of tissue. Therefore, the biomaterial's performances can be different based on the microenvironment that the biomaterial is implanted in. Accordingly, to create harmony and constructive interaction between implantable devices and living tissues in a bioelectricity system, matching the properties of each component is vital [1].

Based on this, the need for new materials to extract data through advanced, immediate and accurate methods has developed different types of materials and methods to improve the interaction of implantable equipment with the biological organs, tissues and cells. Todays, developing of the new high efficient neural interfaces is progressing

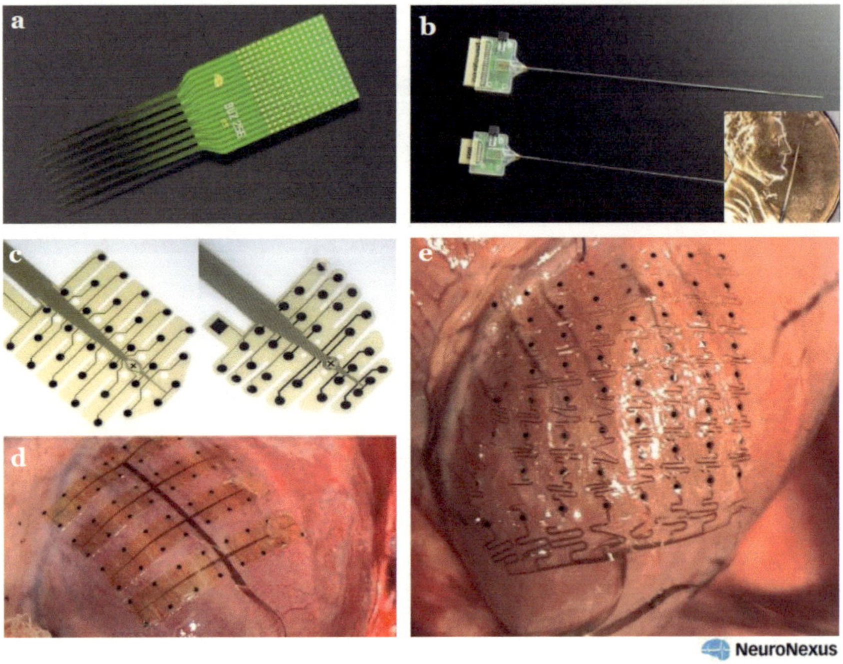

Figure 1.
Different commercial neural electrodes: (a) an iridium (or platinum, gold) multi-arrays electrode fabricated using state-of-the-art silicon microelectromechanical systems (MEMs). (b) a vector array made of silicon (inset: optical images of the electrode size in comparison with a coin), and (c) ultra-flexible Pt electrodes on a polyimide substrate. (d, e) Cardiac surface grid on heart surface. All images are adapted with permission from products of the NeuroNexus company (USA).

rapidly, which is mostly due to the development of new materials. Meanwhile, the current electrodes are usually made of the metals such as platinum (Pt), iridium (Ir) and other materials such as stainless steel and nickel-chromium (Ni-Cr) alloys. **Figure 1** shows various commercial electrodes that are currently used in clinical applications. However, these materials have significant shortcomings such as mechanical mismatch, low biocompatibility, and weak electrochemical performances. Therefore, various new advanced materials such as hydrogels, conductive polymers, and hybrid composites have been developed, so that they are expected to cause the great improvements in bioelectronics and online health monitoring.

2. Essential considerations

2.1 Dynamics and mechanical properties

Reducing the damage to the biological tissue and minimizing the interfering effects on the function of the organ are fundamental considerations in the design and implantation of a device in the body. From the muscle dynamics caused by heart and lung activities to peristalsis of the digestive system and the mobility of bones and joints, may affect the performance of the implant. On the other hand, the presence of biomedical implants in the body can slow down or disrupt the activity of dynamic organs. The forces caused by continuous movements around the implant can easily lead to the destruction of the bioelectronic device. In addition, the stiffness of the living tissue can also affect the performance of the implant. Most biological tissues, except bones and teeth, are soft and have an elastic modulus of less than 1 kPa (such as central nervous system) to higher than 100 kPa (such as lungs, kidneys, arteries) [2]. Meanwhile, pathological conditions can lead to the transformation and increase of stiffness of biological tissues [3]. Subsequently, changing the ECM's stiffness affects the behavior and function of the cells. Therefore, the soft nature of most biological organs and tissues makes them highly vulnerable when faced with biological implants. Accordingly, it is important to understand the dynamics and mechanics of the living tissues.

2.2 Immune responses to implants

After implantation, a layer of proteins, such as fibronectins, collagens, laminins, are quickly absorbed on the implant's surface [4]. This layer is determined by the immune system, leading to foreign body responses (FBR) [5]. The FBR is a phenomenon in which many immune cells undergo apoptosis. This compromises the immune system that may result in formation of a biofilm around the implant and tissue infection [6]. During this process, the formed fibrotic capsule around the implant separates the electronic device from the biological target.

The immune system reactions are altered based on the tissue. In the central nervous system, which is resistant to foreign antigens and immune response [7], the FBR causes the glial scar formation. The resulting scar isolates the bioelectric implant and ultimately impairs the electrical recording/stimulation function. Infections and glial scars caused by the presence of an implant in the body are generally attributed to various reasons such as the mechanical mismatch of the implant with the host tissue [8]. Hence, tailoring the mechanical properties to minimize the FBR is a major consideration when designing an efficient implant. **Figure 2** shows the immune responses, including glial encapsulation and tissue/electrode separation processes, to the implants.

2.3 Chemical environment

The chemical microenvironment surrounding the implant plays an essential role in the stabilization, interaction and performance of the bioelectricity system. Biological tissues are moist and somewhat alkaline microenvironment, which may be aggressive to electronics system components such as sulfate, chloride, carbonate, and phosphate [10]. The combination of the chemical microenvironment with reactive oxygen species (ROS), which are produced during intracellular metabolic processes, can cause the FBR caused by oxidative processes. This disrupts the implant's interaction with the biological tissue, and the subsequent entire functioning of the bioelectricity system.

Ideal biomedical implants should be compatible with the underlying living tissue. This adaptation results in a sincere but not restrictive interaction with the organ's topography. Therefore, to minimize inflammation of the immune system, bioelectronic materials and designs must be biocompatible, and neutral to proteins absorption and immune cells stimulation. For example, applying an inert encapsulating material aims to reduce the direct interaction zone between the electrode and biological tissue, reducing electrical noise and potential electrode degradation. Hence, one of the most effective ways to reduce inflammation and infection, is designing new materials and modifying the implant's surface to be more compatible with biological microenvironments.

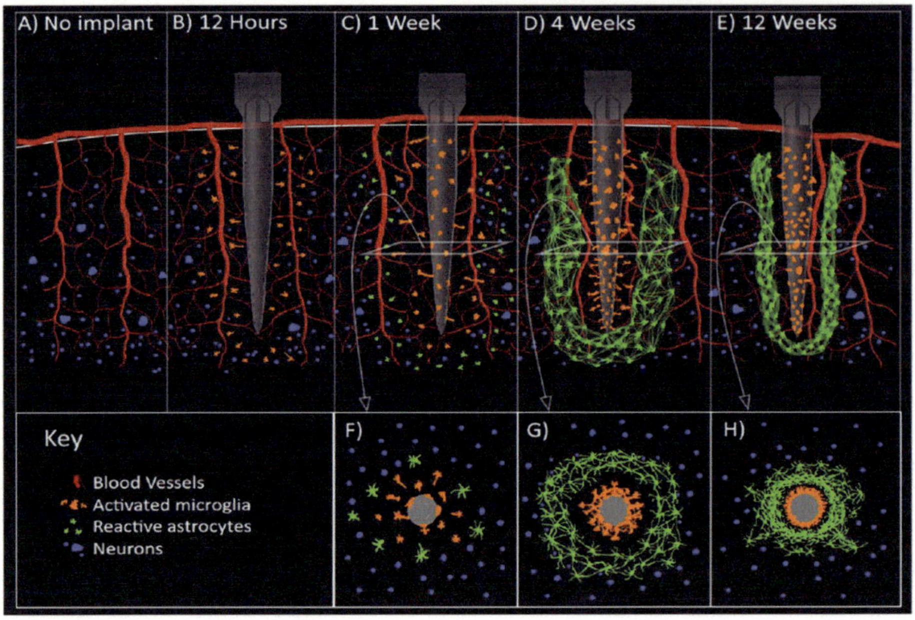

Figure 2.
Glial encapsulation process (A) before implantation, and (B) 12 h post-implantation, (C) 1 week post-implantation, (D) 4 weeks post-implantation, (E) 12 weeks post-implantation. Panels (F), (G), and (H) represent cross-sectional of (C), (D), and (E), respectively. The figure is adapted with permission from Ref. [9] MDPI.

3. Charge transfer mechanisms

3.1 Charge transfer at tissue-biomaterial interfaces

Accurate monitoring of the electrophysiological signals produced by the nervous system and operative tissue stimulation are important parameters for evaluating the performance of the tissue-electronics interface [11]. The generation and transmission of electrophysiological signals are conducted by stimulating ions to pass through the cell's membrane and changing the charge concentration in the ECM. These ions carry the charges in the ECM, so their mobility makes a local electric field. The goal of a neural electrode is to create a two-way connection between the electronics device and biological tissue to record the changes of the electrophysiological field (signal recording) or change the field (electrical stimulation). Enhancing this two-way communication is the basis of that we consider for designing and construction of bioelectricity systems [12]. **Figure 3** indicates the bioelectronics activities in the tissue-electronics interfaces and equivalent circuit models for stimulation and recording. It does not matter which communication mode (recording or stimulation) is performing at the interface, the important thing is that the applied electroactive materials ensure efficient charge transfer between electrons and ions in the bioelectricity system. Because, charge transfer in neural interfaces is done by electrons, while biological tissues transfer electrophysiological charges through ions [14].

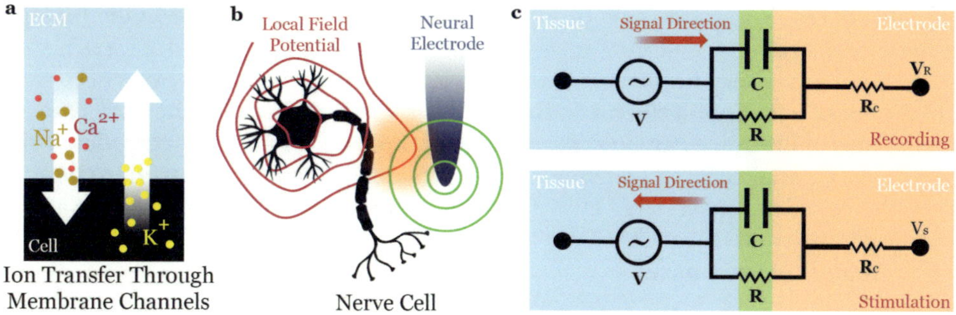

Figure 3.
schematic representation of (a) ions transfer through membrane channels, making (b) an extracellular potential that interacts with the electrical field produced by electrode. (c) Equivalent circuit models for (top) recording and (bottom) stimulation processes. V_R shows the output of the recording with the interconnect resistance R_c. The V indicates the electric potential within ECM generated by the ionic currents applied on the electrode surface. For both recording and stimulation, the tissue-electrode interface is considered as a parallel circuit of the EDL capacitance (C), and the leakage resistance, R. V_S indicates the input for stimulation with the interconnect resistance R_c. The V_e represents the electric potential within ECM generated by V_S applied on the neuron's outer membrane [13].

In a way known as capacitive charge transfer, the transfer mechanism depends on the charging/discharging process of the capacitance made by the electric double layer (EDL) formed on the electrode surface. Upon the electrode generates an electrical pulse, the electrostatic charges concentration on the implant's surface changes, which is accompanied with the alternating absorption and repulsion of ions in the ECM surrounding the implant. It is worth noting that during this process, there is no electron transfer between biomaterials and tissue. In fact, under these conditions, a layer of polarized water molecules is adsorbed on the electrode's surface and acts as a dielectric for the EDL capacitor [15]. It means, no chemicals are produced or consumed in capacitive charge transfer. Therefore, electrodes whose function is based on capacitive charge transfer are more suitable for stable interaction with physiological environments. However, the charge transfer ability of a neural interface depends on the capacitance of the surface EDL, and the capacitance is positively dependent on the surface area of the electrode. Therefore, increasing the effective surface area of neural electrodes, without increasing their size, leads to improvement of the bioelectronics systems performance.

In another charge transfer process that is based on the Faraday mechanism, charge transfer depends on chemical reactions occurring on the electrode's surface. In other words, when the electrode generates an electrical pulse, the reduction-oxidation (redox) reactions occur at the tissue-electrode interface. The Faradaic charge transfer is associated with the reduction or oxidation of chemicals, which establishes as electrons pass through the interface. According to whether new stable products are produced during charge transfer or not, the Faradaic process is divided into two types of reversible or irreversible. During the irreversible Faradaic process, the redox reactions not only lead to the collapse of the electrode, but also damage the living tissue by changing the ECM's conditions such as pH and releasing harmful products into the tissue microenvironment. So, it is preferred to avoid charge transfer processes by irreversible Faradaic paths. Meanwhile, in reversible Faraday charge transfer mechanism, new materials produced on the electrode's surface are converted to their original state during the opposite electric pulse. Hence, in the reversible Faradaic process, new products are not introduced into the living biological tissue. Accordingly, charge transfer through reversible Faradaic processes is safe and desirable. It is very important that the corresponding redox reactions occur during charge injection in the reversible Faradaic method, because they indicate that the neural interface can hold more charges. Consequently, the electrodes with a reversible Faradaic charge transfer mechanism are preferred compared to the electrodes based on irreversible charge transfer. **Figure 4** indicates capacitive and faradaic charge transfer mechanisms at tissue-electronics interfaces and their cyclic voltammetry (CV) responses.

3.2 Charge transfer in biomaterials (electrode)

Due to the type mismatch of charge carriers in the biological tissues and electronics systems, the stable and effective data transmission using the neural interface is a significant challenge. So far, we have discussed charge transfer mechanisms in the biological tissues and the tissue-electrode interfaces. Next, we will describe charge transfer mechanisms in biomaterials (electrode material). Charge transfer in the electrode materials are divided into three main categories, which are:

1. Charge transfer by electrons: This type of charge transfer is often observed in traditional electrodes, which are often made of metals and carbon materials. These materials generally use free electrons as charge carriers to establish connections with biological tissues [16]. Due to the high electrical conductivity and long-term biological stability of metal and carbon electrodes, these materials have been widely used in construction of the neural electrodes [17].

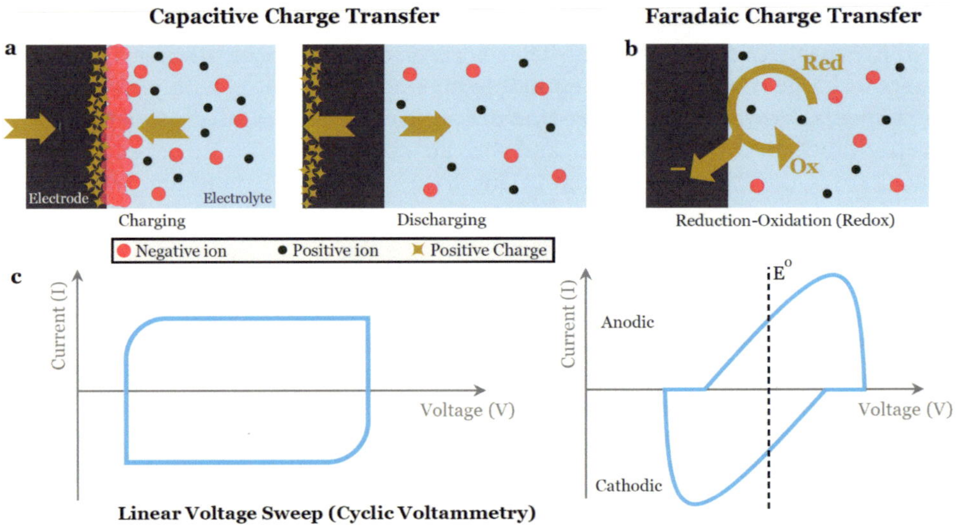

Figure 4.
Schematic illustrations of the electrochemical processes at tissue-electronics interfaces. (a) Capacitive charge transfer based on the electrical double layer (EDL) consisting of the charge and discharge processes. (b) An idealized faradaic (Redox) charge transfer. (c) The capacitive charge transfer characterized by voltage (V)-independent current (I) in cyclic voltammetry, resulting in a box-like cyclic voltammogram. (d) The faradaic charge transfer features an anodic peak (corresponding to oxidation, [Red – e^-→ Ox], where 'Red' and 'Ox' are the reduced and oxidized forms of a species, respectively) and a cathodic peak (corresponding to reduction Ox + e^-→ Red) around the standard redox potential E^o of the redox couple in cyclic voltammetry.

2. Charge transfer by ions: Ion-conducting materials such as hydrogels often have inherent biocompatibility, flexibility and excellent adaptability to the living tissues, which are considered as promising characteristics for the neural interfaces [13, 18, 19].

3. Charge transfer through hybrid electron-ion transfer: This transferring method is often done by conductive polymers (CPs), and their composites with hydrogels and electron conductive materials. In addition to hybrid electron-ion charge transfer mechanism, the excellent mechanical biocompatibility of the CPs has drawn this materials more attention [20].

In the next section, we will describe the notable electrode materials based on their charge transfer mechanisms.

4. Current developing neural interfaces

Tremendous production progresses of new materials, reducing dimensions while increasing efficiency of bioelectronics systems. In addition, the mechanical adaptability and improving the electrical properties of neural electrodes to better interact with biological tissues have always been on the research programs. The improvements of the conventional electrode's properties and design of new electrodes, lead the bioelectronics present in various applications such as heart pacemakers, deep brain stimulators, retinas, contact lens, electronics skin and etc. In this section, we will introduce different types of these electrode materials, manufacturing processes and their functional mechanisms.

4.1 Neural interfaces that transfer charge by electrons

4.1.1 Metal electrodes

Until now, most neural electrodes are mainly produced by electron conductive materials, namely metals and metal composites such as platinum (Pt), gold (Au), silver (Ag), and iridium. However, the practical applications of these materials as electronic-tissue interfaces are limited due to their weak biocompatibility and insufficient electrical activities. It should be noted that the electrical activity of the electrode depends on different electrochemical parameters such as electrochemical impedance, charge injection limit (CIL) and charge storage capacity (CSC). It means, the high electrical conductivity of metal electrodes does not necessarily mean in that they have good electrical activity. For example, it has been demonstrated that although some conductive polymers suffer from lower electrical conductivity, they show higher electrical activity than platinum [21]. Therefore, a material with low electrical conductivity is not necessarily a weak electroactive material, although increasing the electrode's conductivity usually improves its electrical activity [22]. In the following, we will discuss the recent approaches for improving the biological behaviors and electrical activities of the metal electrodes.

- Nanostructured metal electrodes

Do more with less, this is the mantra of nanotechnology. Considering the urgent need for small bioelectricity systems, increasing the number of channels and charge injection density of neural electrodes while reducing their dimensions, researchers have developed different types of electrodes with nanoscale features [23, 24]. However, these electrodes still face unavoidable limitations, which are: (I) the possibility of bending and placing in unspecified places due to the excessive fineness and frangibility. (II) To design these tiny electrodes, the materials with high hardness and stiffness are needed, which may cause activation of the immune system, formation of glial scars and tissue inflammation. (III) Nanometer electrodes may have relatively high electrochemical impedance and low charge injection capability, which leads to weakening the performance of the bioelectricity system. However, some approaches such as introducing nanopores in the electrode structure for increasing its effective surface area, can reduce the electrochemical impedance of the neural interface [25]. Recently, composite coatings made of electrodeposited iridium oxides with Pt gray were developed to fabricate IrO_x/Pt gray neural electrode. It demonstrated that the large surface area of the nanoporous Pt, leading to firm adhesion of the iridium oxide to the substrate accompanying with superior mechanical and electrochemical stability of the electrode [26]. Other example are the diamond-titanium porous composites produced by deposition process to fabricate hybrid neural electrodes. These composite electrodes have demonstrated high electrochemical capacitance, low impedance, and excellent biocompatibility assessed in vitro using cortical neurons [27]. Moreover, the nanostructured electrodes have shown greater compatibility with the ECM [28]. The flexibility of nanostructured electrodes in line with the weak movements of the brain and other dynamic biological organs, is another advantage of nanometer electrodes [29].

- Metal composite electrodes

High impedance and low biocompatibility are the unfavorable characteristics of traditional metal electrodes for monitoring electrophysiological signals and tissue stimulation. Accordingly, metal composites have been developed to overcome the mentioned challenges and improve the long-term stability of metal-based neural interfaces. Avoiding parasitic effects should be considered for design of metal composites, so combining the materials with similar characteristics is more preferable. In addition, it has been proved that the nanocomposites, which have high specific surface area and more compatibility with the ECM, improve the performance of the neural interface. Also, fabrication of nanostructured composite coatings on the electrode, could be a method to create the preferred properties. The composite coatings are usually deposited by electrochemical deposition, sputtering and thermal evaporation methods. Currently, gold (Au) is considered as a promising candidate for improving the properties of neural electrodes. The excellent biocompatibility and encouraged performance of the Au-coated electrodes have been proven [30]. Compared to the pure silver (Ag) surface, the Au-Ag nanocomposite electrodes have shown lower impedance and more biocompatibility, which has resulted in accurate, high-quality, and stable recording of electrocardiogram (ECG) and electromyogram (EMG) signals. Moreover, the traditional neural electrodes generate wide signal void (no functional magnetic resonance imaging (MRI) signal) in ultrahigh field (UHF) MRI scanners. This is an important shortcoming when simultaneous MRI signal acquisition and neural monitoring is desired, for example in studying the functional mechanisms of deep brain stimulation (DBS). Recently, new gold-aluminum (Au-Al) composites have been presented for neural interfaces to overcome the signal voids. The Au-Al composites significantly reduce the magnetic susceptibility difference between the brain tissue and electrode, resulting in greatly reduced regions of the signal voids. The Au-Al composites produced less field distortion and signal loss compared to the pure Au and Al electrodes, leading to MRI scanners of lower magnetic field strengths as well [31].

Despite to the conventional electrode metals such as the toxic Ag nanowires [32], liquid metals (LM) have good biocompatibility, excellent mechanical flexibility, and significant electrical conductivity. The LM composites include bismuth-indium-tin (Bi-In-Sn), indium-gallium eutectic (EGaIn) and gallium-indium-tin (GaInSn), are widely used in the preparation of neural electrodes. However, these LM electrodes have obvious challenges, including (I) removing the oxide layer of LM particles inside the ink to connect the electrode paths, (II) overcoming the surface tension of the LM and converting it into desired shape, (III) solving the problem of connection fragility of the LM to other components of the bioelectricity system, and (IV) encapsulating the LM to prevent leakage into the biological microenvironment, and subsequent tissue damages.

4.1.2 Carbon-based electrodes

At present, carbon-based materials such as carbon nanotubes (CNTs) and graphene, are outstanding candidates for the design and fabrication of nanoscale, flexible, and multifunctional neural interfaces.

- Graphene electrodes

Graphene is a single layer of carbon atoms that are connected to each other in the form of a two-dimensional honeycomb network by sp^2 hybridization. Recently, graphene has been used as a successful high efficiency material in the neural interfaces [33]. Among the prominent features of the graphene, the following should be mentioned: (1) very high mechanical strength while maintaining extraordinary flexibility, (2) excellent electrical conductivity (single layer resistance of 100 Ω $sq.^{-1}$) with a carrier mobility of ~15,000 $cm^2\ V^{-1}\ s^{-1}$ at ambient temperature [34], (3) large surface area that provides a favorable template for cell attachment and charge transfer, thereby enhancing electrical recording/stimulation capability, and (4) biocompatibility. Additionally, the modified surface functional groups of the graphene, makes it more compatible as an operational nanostructure for various applications. By manipulating or functionalizing graphene, some modified structures such as graphene oxide (GO), reduced graphene oxide (rGO), graphene fibers and other derivatives can be produced, which brings more choices for bionanotechnology [35, 36]. Graphene can be coated on the electrodes using different methods such as chemical vapor deposition (CVD) [37], spraying [38], and electrochemical routes such as cyclic voltammetry (CV) [39]. However, weak adhesion of the graphene to the substrate, resulting in coating instability and destruction of the electrode, is one of the main challenges of the graphene electrodes.

- CNT electrodes

Carbon nanotubes (CNTs) are one-dimensional nanostructures created by twisting the graphene sheets. The CNTs have been widely used in the construction of neural electrodes, duet to their following desirable characteristics: (1) they have high electrical conductivity, facilitating the ballistic electron transfer from the electronics-tissue interface to the electrode material. (2) They have a high surface-to-volume ratio, which can reduce the electrochemical impedance and increase the charge injection capability of the electrode [40, 41]. (3) The CNTs benefit from surface functional groups that are easily modified by biological molecules, leading to tunable anisotropic properties adequate based on the application. In addition, (4) The CNT-based electrodes have presented superior biocompatibility, mechanical strength, flexibility, and worthy adhesion to the nerve cells [42, 43].

There is a trade-off between the size of the neural electrode and its electrochemical impedance, so that by decreasing the electrode size decreases, its impedance increases. It should be noted that for high resolution and accurate recording of the electrophysiological signals, the small electrodes with low impedance are needed. Although the size and electrochemical properties of neural electrodes have been continuously improved, weakening the mechanical performances during miniaturization have always been an important challenge for practical applications [44]. Therefore, the CNTs are suitable nanomaterials to overcome this challenge. In addition, the improvement of biocompatibility, reduction of impedance, promotion of more stable micro-environmental ability of the conventional electrodes modified with the CNTs have also been reported [45]. The researchers have evaluated conductive CNTs/collagen composites for studying the cellular responses on the neural interfaces. The results indicated that by increasing the collagen content, the cells show enhanced attachment on the electrode's surface, which could be due to the high ability of the collagen to improve the adhesion and viability of the nerve cells [46].

4.2 Neural interfaces that transfer charge by ions

Signal transmission in biological media is conducted through the movement of ions and small molecules, contrary to the electron-hole in electronic devices. Accordingly, the ion-conducting neural interfaces can interact more efficient with the living tissue. The finding this tissue-electronics charge transfer interaction has caused the ion-based interfaces to receive more attentions. However, the use of liquid ion-conducting materials such as electrolyte solutions and ionic liquids, which have high ion transfer capability, are limited due to the need for a mold to maintain the shape of the electrode [47]. As a result, the solid ion conductors have attracted more attentions. Recently, hydrogels have been broadly applied in biological applications, including cell culture, smart drug delivery, tissue repair and regeneration, due to their intrinsic biocompatibility, biological functionality, flexibility, and adaptation to living nerve tissue. The outstanding capabilities of the hydrogels have caused these materials to be considered promising candidates for designing and manufacturing flexible bioelectronic systems [48].

4.2.1 Hydrogels

Hydrogels are soft solid ion-conducting materials that are composed of interconnected polymeric configuration. This polymeric structure can absorb water that enables free ion movement in aqueous-based network, creating ionic conductivity. This ability has caused hydrogels to be used in various applications such as artificial muscles, ionic skin, artificial axons and connections of the central nervous system, whose functions are performed through ion conduction. Conductive hydrogels are usually synthesized by (1) constructing distinct component networks through

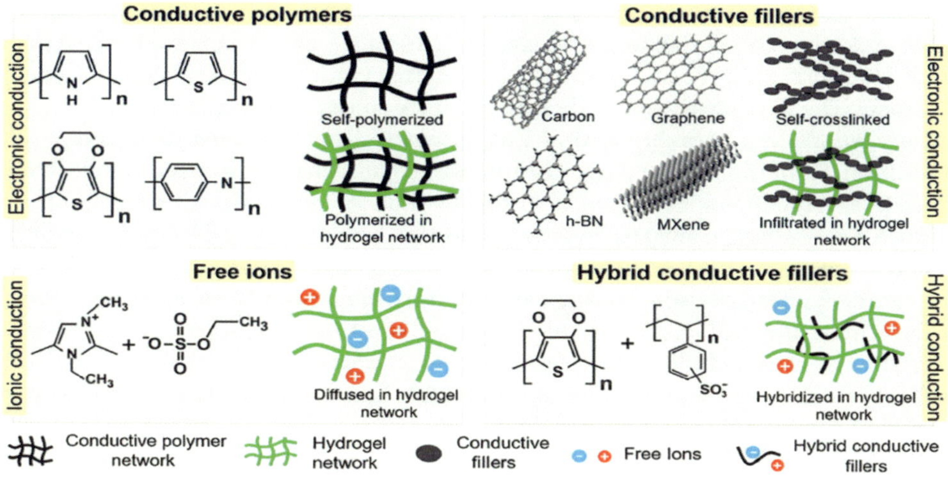

Figure 5.
Different types of conductive hydrogels, preparing by using CPs, conductive fillers, free ions, and their composites. According to the composite's materials and properties, the fabricated hydrogels can be classified as electronic, ionic, and the hybrid electron–ion conducting hydrogels. The figure is adapted with permission from Ref. [49] American Chemical Society.

self-assembly or self-polymerization CPs/fillers, (2) building interpenetrating networks by doping CPs/fillers, (3) diffusing free ions, and (4) embedding conductive fillers/free ions into an existing non-conductive hydrogel matrix (**Figure 5**). Depending on the additives and dominant conducting mechanisms, the synthesized hydrogels can be classified as electron-based, ion-conducting, and hybrid electron–ion conductive hydrogels.

The ionic conduction is mainly based on a non-Faradaic (capacitive) charge transfer mechanism, without materials or charges passing through the neural interface. This transmission possesses the transmitting high-frequency electrical signals over long distances [50].

There are three types of widely used hydrogels for the electronic-tissue interfaces, which are (1) ion-conducting hydrogels (ICHs), (2) ion-conducting organohydrogels (ICOHs), and (3) hydrogel composites. In the following, we will focus on hydrogels in which the electrical conduction is conducted only through ions. Other hydrogels in which the charge transfer process is carried out by hybrid electron-ion transfer will be discussed in the next section.

The ICHs are classified as highly hydrophilic gels with a massive three-dimensional (3D) hydrated network structure [51]. Since ions can move freely in this 3D network, ICHs can achieve high ionic conductivity in the range of 3.4–5.5 Sm^{-1} [52, 53] by permeating salts such as NaCl, LiCl, $FeCl_3$, acids such as HCl, H_3PO_4, or ionic liquids [54–56]. In addition, the ion conduction mechanism of hydrogels is similar to biological tissues, so they can efficiently exchange data through ion diffusion [48]. This makes hydrogels immune to the challenges of converting electronics and ion-based signals to each other and related problems [50, 57]. It should be noted that this group of hydrogels contain large amounts of water, which can act as a buffer environment during the side effects, to prevent adverse problems for the living tissue. Most ICHs are adhesive, transparent, and self-healing, so their applications in bioelectronic systems such as wearable sensors, implantable epidermal electrodes, digital tattoos, and many others are being developed [58, 59].

However, the prepared ICHs by salt compounds, have unfavorable biocompatibility and low stability due to the release of excess ions, which can lead to the damage of bioelectronic devices. It has been proven that compared to traditional metal electrodes, the ICH-based electrodes can generate contractile forces using lower voltages, which indicates the capability of the electrodes for bioelectrical stimulation. In addition, it should be noted that the undesired gaps formed between the electronic interface and the biological tissue, resulting by muscle contraction or skin bending (surface electromyography abbreviated to sEMG), leads to significant noise/error in the identification of electrophysiological signals. These gaps can be eliminated using soft hydrogel interfaces and making electrostatic interactions between the electrode's surface and tissue [60]. However, the ICH-based electrodes lose conductivity, flexibility, and even morphology due to rapid water loss, which limits their practical applications [61]. At present, some solutions have been proposed to overcome this problem, including (I) addition of dehydrating reagents [62, 63], (II) mixing of ionic-polymeric liquid gels [59], (III) binding of sealing materials [64], and (IV) mixing of deep eutectic solvents (DESs).

As mentioned, the water losing is one of the major shortcomings of hydrogels in practical applications. Recently, adding organic solvents to hydrogels for producing the ion-conducting organohydrogels (ICOHs) has been considered to overcome the dehydration of hydrogels. This modification is based on the premise that the organic solvents can compensate some of the lost water of hydrogels, and enhance their dry immunity and maintain ionic conductivity [65]. In addition, the ICOHs retain some advantages of the hydrogels, including biocompatibility, soft mechanical properties, and considerable shape design ability [66, 67]. Adding organic solvents to hydrogels can be conducted through three methods, which are (1) solvent replacement [68]; (2) the desorbed hydrogel network is injected with organogel precursors, and then is subjected to in situ polymerization [69]; and (3) gelation in a binary solvent [70].

It is worth noting that although the solvent displacement method could be done easily, the ICOHs prepared by this method have relatively weak forces between the hydrogel's polymer network and the replaced solution, resulting in solvent leakage and tissue damages [71]. Meanwhile, ICOHs synthesized by the gelation method in binary solvents have overcome this challenge and also presented the advantage of high electrical conductivity. While, high electrochemical impedance and insufficient long-term adhesion to biological tissue are some problems of the ICOHs, which have limited their usage in the bioelectronics.

4.3 Neural interfaces with hybrid electron-ion charge transfer mechanism

An ideal tissue-electronics interface should provide the charge transfer requirements of biological tissues and electronics devices simultaneously. Hence, the materials with hybrid electron-ion charge transfer are more appropriate for design and construction of the neural interfaces. Recently, conducting polymers (CPs) have attracted many attentions to create neural interfaces due to their hybrid electron-ion charge transfer capability.

4.3.1 Conductive polymers

The CPs benefit from unique features, which include (1) simultaneous electron-ion conductivity [72], leading to reduction of the electrochemical impedance and improvement of the electrical recording/stimulation. (2) Their inherent adaptive mechanical properties lead to bridging at electronic-tissue interfaces. (3) Fibers and nanostructures of the CPs, which could be synthesized by low-cost and simple methods, have high specific surface areas that result in facilitating electron-ion exchange at neural electronics-tissue interfaces. (4) Mixing the CPs with different materials such as polyelectrolytes (polystyrene sulfonic acid, polyacrylic acid, polymethacrylic acid) and bioactive molecules [73] can be easily conducted, which improves the biocompatibility and stability of the CPs-based electrodes [74]. In the following, some well-known CPs materials, which have shown promising performances as the neural electrodes, will be presented.

- Neural electrodes made from polypyrrole

Polypyrrole (PPy) is an intrinsic CP that has high conductivity, biocompatibility, facile processing, water solubility, and slight potential for its monomers oxidation, which has made it a capable candidate for the neural interfaces [75]. An ideal CP should have independent decent properties and performances as much as possible, without adding extra reinforcements. To achieve this goal, the CPs need to have mechanical and chemical stabilities in biological microenvironments. However, contrary to the mentioned worthy properties of the PPy, it is prone to irreversible oxidation and easily fails under the change of chemical conditions such as pH and disrupts the bioelectronics interfaces. This characteristic has limited the PPy applications, resulting in more attentions to alternative CPs such as the poly (3,4-ethylenedioxythiophene) (PEDOT).

- PEDOT electrodes

The PEDOT is an intrinsically CP that appears to be more attractive than PPy due to several reasons, including (1) PEDOT has a narrow band gap [76], changing charges in the polymer chains. (2) PEDOT has a higher electrical conductivity that increases the electrode's capabilities [22, 77]. (3) PEDOT has shown great electrochemical stabilities [78], which are necessary for the electrical recording/excitation stabilization in a bidirectional communication. In addition, (4) the PEDOT has better biocompatibility than PPy [79, 80]. Although the CPs have many advantages in bioelectronics, their brittleness and excessive stiffness have limited their use in neural electrodes. Therefore, the design and construction of soft and elastic composites based on conductive elastomers (such as PEDOT:PSS) distributed in a soft elastomer matrix (such as polyurethane, PU), and or using the laser micromachining technology for converting the CPs into a flexible electrodes array, have been proposed [81].

- CPs composite electrodes

In addition to the CPs-based electrodes, surface modification of traditional electrodes using the CPs and their composites has given innovative capabilities to the electrical recording/stimulation processes. Improving the performance of electrodes with the CPs-based coatings, could be due to (1) reducing the electrochemical impedance of the electrode, (2) making a soft compatible surface for the electrode while being strong to improve the tissue-electronics interface. In addition, (3) increasing bioactivity in comparison with the bio-inert metallic electrodes, results in decreasing immune responses, tissue inflammation and implant infections. The CPs not only increase the electrode's stability, but also their highly porous structures, such as electrospun fibers, can improve the electrochemical performances [82]. The biological fluids can flow in the pores of fibrous structures and interact with a high surface area to increase the electrophysiological signal transmission efficiency. Additionally, making porous composites such as fibrous CPs/CNTs can also enhance the polymer's conductivity and increase the electrode's effective surface area. The results have also shown that the neural interfaces based on the CPs/CNTs composites have lower impedance and more charge storage capabilities. These improvements can be due to the presence of CNTs as impurities to make a strong interaction with the CP chains, leading to fast electron transfer processes, formation of three-dimensional structures, and increasing the effective surface of the composite electrode [83].

- Neural electrodes made from hydrogel composites

Hydrogels can make ionic conductivity through the absorption and transfer of the ions present in their structural water, which is a process similar to the transfer of electrophysiological signals. Although this hydrogel's property has made them attractive, activation electron transfer mechanisms is the key for successful bioelectronics application of the hydrogels [84, 85]. Fortunately, the hydrogel's porous structure provides adequate space to incorporate with the electron-conducting materials such as metals, carbon-based materials, conductive polymers, etc. with the aim of forming a composite of hybrid electron-ion transfer network. Accordingly, it is possible to improve the electrochemical properties of the hydrogels, including electrochemical impedance and CIL, without weakening their excellent biological properties [49, 86].

The CPs are ideal materials for making hydrogel-based composites, because (1) the similar soft and high flexible mechanical properties of the CPs and hydrogels, avoiding mismatch dynamics and extra stresses to the electrode. (2) The CPs have high compatibility and affinity to the hydrogels that results in potential hybridization. In addition, (3) the unique polymeric and organic properties of the CPs make them facile to be modified. It should be noted that (4) some CPs such as PEDOT have a hydrogel-like form in wet environments, which corresponded to their hygroscopicity or swelling properties when exposed to water [13]. This behavior makes this kind of CPs more similar to the hydrogels' properties, leading to more integrated and consistent composites.

The CPs are usually used as the electron conducting additives in hydrogel composites. The polymeric nature of CPs allows the molecular scale structures, forming interpenetrating network (IPN) hydrogels. The IPNs not only minimize potential trade-offs in mechanical properties, but also meaningfully reduce probable heterogeneity in electrical and mechanical behaviors. The IPN-based hydrogels could be generally synthesized via three main approaches, including (a) direct mixing of CPs and the hydrogel's precursors, (b) in-situ polymerization on CPs in the hydrogel matrix, and (c) in-growth polymerization of CPs into the hydrogel matrixes. **Figure 6** shows schematic illustrations of the IPNs production processes.

Making nanocomposite using electron-conducting nanostructures, including carbon nanotubes, liquid metals, graphene, metal nanowires, is another approach to increase the hydrogels' conductivity. Carbon-based materials are preferred candidates for mixing with hydrogels, because their excellent conductivity, high effective surface area, good chemical stability in wet environments, and especially the ability to form covalent bonds with different polymeric network groups. In addition, under the premise of satisfying improved biocompatibility and electrochemical behavior,

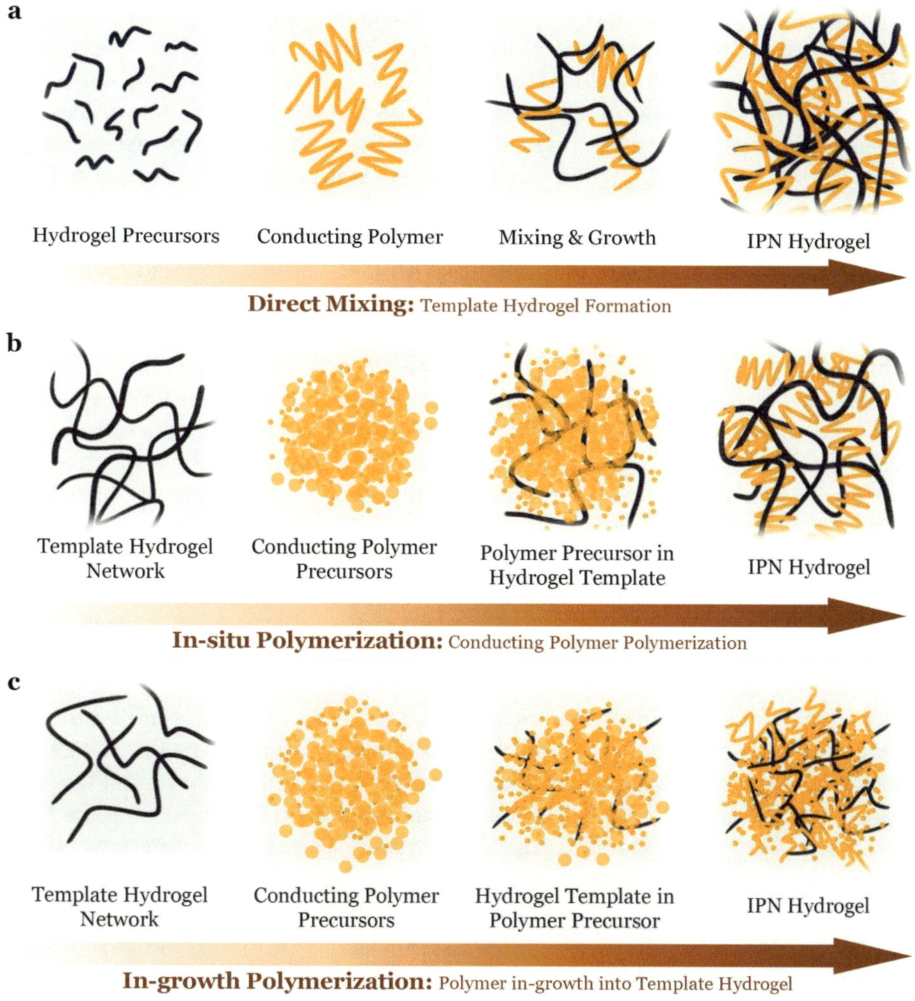

Figure 6.
The main synthesis methods towards the CPs IPN hydrogels. (a) direct mixing of CPs and the hydrogel's precursors, (b) in-situ polymerization on CPs in the hydrogel matrix, and (c) in-growth polymerization of CPs into the hydrogel matrixes [13].

the bioelectronics interfaces created by the materials with hybrid electron-ion charge transferring, shows more efficient signal transmission between the biological tissues and the external bioelectronics systems.

5. Conclusion and outlook

Tissue-electronics interfaces are the key component of bioelectronics systems. These interfaces provide a two-way communication between biological tissue and external electronics devices to record electrophysiological signals and stimulation of the living organs. In biological tissues, the charge transfer is done through ions, while the electronic systems generally use the electrons as electric charge carriers.

Accordingly, the neural interfaces with a hybrid electron-ion charge transfer mechanism can improve the charge transfer processes at bioelectronics. There are two main charge transfer mechanisms at the tissue-electronic interfaces, including capacitive (based on EDL) and Faraday charge transfer. In addition, the biocompatibility, which can reduce FBR and glial scar formation, is an essential parameter for tissue-electrode adhesion improvement that results in accurate and stable signal transmission. Recent general methods to increase the biocompatibility of neural interfaces include reducing the biomaterial's stiffness and elastic modulus, biocompatible coatings, and developing new bioactive materials and composites. Although significant progress has been made in the design and fabrication of tissue-electronic interfaces, some challenges must be overcome before the bioelectronics interfaces can be efficiently operated, including (I) the weak adhesion of the neural electrode materials to biological tissues. This problem may be due to the high hydrophobicity and insufficient biocompatibility of the current interfaces. (II) Inadequate biocompatibility of the electrode materials that leads to FBR, glial scar formation and subsequent tissue-electronics separation. The challenge is that although the electrode materials should have low stiffness and elastic modulus to avoid mechanical mismatch with host living tissue, the weak mechanical properties, especially in dynamic organs, can lead to electrode destruction and failure. Therefore, adapting the properties of the implant to the host living tissue, is important for successful implantation and long-term stability of the electrode. The efficiency of the neural recording/stimulation process depends on the biocompatibility, electrochemical properties such as impedance, charge storage capability, and charge injection limitation of the electrode. High electrochemical impedance reduces the accuracy of electrophysiological signals recording. Reducing methods of the electrochemical impedance mainly include the following: (1) enhancing the conformal capability between the neural interfaces, (2) improving the tissue-electronic adhesion, and (3) modifying the electrode's surface characteristics to adapt with the ECM. In summary, the prospect of developing high efficient tissue-electronic interfaces due to rising the new materials and strategies is expected. To achieve this goal, the main neural biomaterials' properties such as biocompatibility and electrochemical performance must first be improved. It is believed that the development of neural interfaces can contribute to the progress of bioelectronics medicine, neuroscience, and online health monitoring.

Conflict of interest

The authors declare no conflict of interest.

Author details

Shahab Ahmadi Seyedkhani* and Raheleh Mohammadpour
Institute for Nanoscience and Nanotechnology (INST), Sharif University of Technology, Tehran, Iran

*Address all correspondence to: sh.ahmadi@sharif.edu

References

[1] Fleischer S, Dvir T. Tissue engineering on the nanoscale: Lessons from the heart. Current Opinion in Biotechnology. 2013;**24**:664-671

[2] Scholten K, Meng E. Materials for microfabricated implantable devices: A review. Lab on a Chip. 2015;**15**:4256-4272

[3] Hiesinger W, Brukman MJ, McCormick RC, Fitzpatrick JR III, Frederick JR, Yang EC, et al. Myocardial tissue elastic properties determined by atomic force microscopy after stromal cell–derived factor 1α angiogenic therapy for acute myocardial infarction in a murine model. The Journal of Thoracic and Cardiovascular Surgery. 2012;**143**:962-966

[4] Rahnamaee SY, Bagheri R, Vossoughi M, Ahmadi Seyedkhani S, Samadikuchaksaraei A. Bioinspired multifunctional TiO2 hierarchical micro/nanostructures with tunable improved bone cell growth and inhibited bacteria adhesion. Ceramics International. 2020;**46**:9669-9679. DOI: 10.1016/j.ceramint.2019.12.234

[5] Wilson CJ, Clegg RE, Leavesley DI, Pearcy MJ. Mediation of biomaterial–cell interactions by adsorbed proteins: A review. Tissue Engineering. 2005;**11**:1-18

[6] Costerton JW, Montanaro L, Arciola CR. Biofilm in implant infections: Its production and regulation. The International Journal of Artificial Organs. 2005;**28**:1062-1068

[7] Pachter JS, de Vries HE, Fabry Z. The blood-brain barrier and its role in immune privilege in the central nervous system. Journal of Neuropathology and Experimental Neurology. 2003;**62**:593-604

[8] Chen R, Canales A, Anikeeva P. Neural recording and modulation technologies. Nature Reviews Materials. 2017;**2**:1-16

[9] Campbell A, Wu C. Chronically implanted intracranial electrodes: Tissue reaction and electrical changes. Micromachines. 2018;**9**:1-14. DOI: 10.3390/mi9090430

[10] Xin Y, Huo K, Tao H, Tang G, Chu PK. Influence of aggressive ions on the degradation behavior of biomedical magnesium alloy in physiological environment. Acta Biomaterialia. 2008;**4**:2008-2015

[11] Kolaya E, Firestein BL. Deep brain stimulation: Challenges at the tissue-electrode interface and current solutions. Biotechnology Progress. 2021;**37**:e3179. DOI: 10.1002/btpr.3179

[12] Cogan SF. Neural stimulation and recording electrodes. Annual Review of Biomedical Engineering. 2008;**10**:275-309

[13] Yuk H, Lu B, Zhao X. Hydrogel bioelectronics. Chemical Society Reviews. 2019;**48**:1642-1667

[14] Harris AR, Wallace GG. Electrochemical methods for analysing and controlling charge transfer at the electrode–tissue interface. Current Opinion in Electrochemistry. 2019;**16**:143-148

[15] Merrill DR, Bikson M, Jefferys JGR. Electrical stimulation of excitable tissue: Design of efficacious and safe protocols. Journal of Neuroscience Methods. 2005;**141**:171-198

[16] Dong R, Wang L, Hang C, Chen Z, Liu X, Zhong L, et al. Printed stretchable

liquid metal electrode arrays for in vivo neural recording. Small. 2021;**17**:2006612

[17] Kim YH, Koo H, Kim MS, Jung S-D. Iridium oxide on indium-tin oxide nanowires: An all metal oxide heterostructured multi-electrode array for neuronal interfacing. Sensors and Actuators B: Chemical. 2018;**273**:718-725

[18] Sung C, Jeon W, Nam KS, Kim Y, Butt H, Park S. Multimaterial and multifunctional neural interfaces: From surface-type and implantable electrodes to fiber-based devices. Journal of Materials Chemistry B. 2020;**8**:6624-6666

[19] Green R. Elastic and conductive hydrogel electrodes. Nature Biomedical Engineering. 2019;**3**:9-10

[20] Zeglio E, Rutz AL, Winkler TE, Malliaras GG, Herland A. Conjugated polymers for assessing and controlling biological functions. Advanced Materials. 2019;**31**:1806712

[21] Balint R, Cassidy NJ, Cartmell SH. Conductive polymers: Towards a smart biomaterial for tissue engineering. Acta Biomaterialia. 2014;**10**:2341-2353

[22] Chen N, Tian L, Patil AC, Peng S, Yang IH, Thakor NV, et al. Neural interfaces engineered via micro- and nanostructured coatings. Nano Today. 2017;**14**:59-83. DOI: 10.1016/j.nantod.2017.04.007

[23] Schwarz DA, Lebedev MA, Hanson TL, Dimitrov DF, Lehew G, Meloy J, et al. Chronic, wireless recordings of large-scale brain activity in freely moving rhesus monkeys. Nature Methods. 2014;**11**:670-676

[24] Riahi Z, Seyedkhani SA, Sadrnezhaad SK. Electrophoretic encapsulation for slow release of vancomycin from perpendicular TiO_2 nanotubes grown on Ti6Al4V electrodes. Materials Research Express. 2020;**6**:125424. DOI: 10.1088/2053-1591/ab6c98

[25] Moxon KA, Hallman S, Aslani A, Kalkhoran NM, Lelkes PI. Bioactive properties of nanostructured porous silicon for enhancing electrode to neuron interfaces. Journal of Biomaterials Science. Polymer Edition. 2007;**18**:1263-1281

[26] Zeng Q, Xia K, Sun B, Yin Y, Wu T, Humayun MS. Electrodeposited iridium oxide on platinum nanocones for improving neural stimulation microelectrodes. Electrochimica Acta. 2017;**237**:152-159. DOI: 10.1016/j.electacta.2017.03.213

[27] Mani N, Ahnood A, Peng D, Tong W, Booth M, Jones A, et al. Single-step fabrication method toward 3d printing composite diamond–titanium interfaces for neural applications. ACS Applied Materials & Interfaces. 2021;**13**:31474-31484. DOI: 10.1021/acsami.1c07318

[28] Rahnamaee SY, Ahmadi Seyedkhani S, Eslami Saed A, Sadrnezhaad SK, Seza A. Bioinspired TiO2/Chitosan/HA coatings on Ti surfaces: Biomedical improvement by intermediate hierarchical films. Biomedical Materials. 2022;**17**:1-17. DOI: 10.1088/1748-605X/ac61fc

[29] Liang C, Liu Y, Lu W, Tian G, Zhao Q, Yang D, et al. Strategies for interface issues and challenges of neural electrodes. Nanoscale. 2022;**14**:3346-3366

[30] Chapman CAR, Chen H, Stamou M, Biener J, Biener MM, Lein PJ, et al. Nanoporous gold as a neural interface coating: Effects of topography, surface chemistry, and feature size. ACS Applied Materials & Interfaces. 2015;**7**:7093-7100

[31] Cruttenden CE, Ahmadi M, Zhang Y, Zhu X-H, Chen W, Rajamani R. Novel composite gold-aluminum electrode with application to neural recording and stimulation in ultrahigh field magnetic resonance imaging scanners. Annals of Biomedical Engineering. 2021;**49**:2337-2348. DOI: 10.1007/s10439-021-02779-y

[32] Theodorou IG, Müller KH, Chen S, Goode AE, Yufit V, Ryan MP, et al. Silver nanowire particle reactivity with human monocyte-derived macrophage cells: Intracellular availability of silver governs their cytotoxicity. ACS Biomaterials Science & Engineering. 2017;**3**:2336-2347

[33] Deng J, Yuk H, Wu J, Varela CE, Chen X, Roche ET, et al. Electrical bioadhesive interface for bioelectronics. Nature Materials. 2021;**20**:229-236

[34] Liu S, Zhao Y, Hao W, Zhang X-D, Ming D. Micro-and nanotechnology for neural electrode-tissue interfaces. Biosensors & Bioelectronics. 2020;**170**:112645

[35] Lu Y, Liu X, Hattori R, Ren C, Zhang X, Komiyama T, et al. Ultralow impedance graphene microelectrodes with high optical transparency for simultaneous deep two-photon imaging in transgenic mice. Advanced Functional Materials. 2018;**28**:1800002

[36] Lu Y, Liu X, Kuzum D. Graphene-based neurotechnologies for advanced neural interfaces. Current Opinion in Biomedical Engineering. 2018;**6**:138-147

[37] Koerbitzer B, Krauss P, Nick C, Yadav S, Schneider JJ, Thielemann C. Graphene electrodes for stimulation of neuronal cells. 2D Materials. 2016;**3**:24004

[38] Wang L-J, Li L, Yu J, Wu Y, He H, Ouyang X, et al. Large-area graphene coating via superhydrophilic-assisted electro-hydrodynamic spraying deposition. Carbon N. Y. 2014;**79**:294-301

[39] Yahya Rahnamaee S, Bagheri R, Vossoughi M, Khafaji M, Asadian E, Seyedkhani SA, et al. A new approach for simultaneously improved osseointegration and antibacterial activity by electrochemical deposition of graphene nanolayers over titania nanotubes. Applied Surface Science. 2022;**580**:152263. DOI: 10.1016/J.APSUSC.2021.152263

[40] Shein M, Greenbaum A, Gabay T, Sorkin R, David-Pur M, Ben-Jacob E, et al. Engineered neuronal circuits shaped and interfaced with carbon nanotube microelectrode arrays. Biomedical Microdevices. 2009;**11**:495-501

[41] Meyyappan M. Carbon Nanotubes: Science and Applications. Boca Raton: CRC Press; 2004

[42] Nguyen-Vu TDB, Chen H, Cassell AM, Andrews RJ, Meyyappan M, Li J. Vertically aligned carbon nanofiber architecture as a multifunctional 3-D neural electrical interface. IEEE Transactions on Biomedical Engineering. 2007;**54**:1121-1128

[43] John AA, Subramanian AP, Vellayappan MV, Balaji A, Mohandas H, Jaganathan SK. Carbon nanotubes and graphene as emerging candidates in neuroregeneration and neurodrug delivery. International Journal of Nanomedicine. 2015;**10**:4267

[44] Li H, Wang J, Fang Y. Bioinspired flexible electronics for seamless neural interfacing and chronic recording. Nanoscale Advances. 2020;**2**:3095-3102

[45] Vafaiee M, Mohammadpour R, Vossoughi M, Asadian E, Janahmadi M, Sasanpour P. Carbon nanotube modified microelectrode array for neural interface.

Frontiers in Bioengineering and Biotechnology. 2021;**8**:582713

[46] Cho Y, R. Ben Borgens, The effect of an electrically conductive carbon nanotube/collagen composite on neurite outgrowth of PC12 cells. Journal of Biomedical Materials Research. Part A. 2010;**95**:510-517. DOI: 10.1002/jbm.a.32841

[47] Wang C, Yokota T, Someya T. Natural biopolymer-based biocompatible conductors for stretchable bioelectronics. Chemical Reviews. 2021;**121**:2109-2146

[48] Sheng H, Wang X, Kong N, Xi W, Yang H, Wu X, et al. Neural interfaces by hydrogels. Extreme Mechanics Letters. 2019;**30**:100510

[49] Fu F, Wang J, Zeng H, Yu J. Functional conductive hydrogels for bioelectronics. ACS Materials Letters. 2020;**2**:1287-1301

[50] Yang C, Suo Z. Hydrogel ionotronics. Nature Reviews Materials. 2018;**3**:125-142

[51] Yang CH, Chen B, Lu JJ, Yang JH, Zhou J, Chen YM, et al. Ionic cable. Extreme Mechanics Letters. 2015;**3**:59-65

[52] Ye Y, Zhang Y, Chen Y, Han X, Jiang F. Cellulose nanofibrils enhanced, strong, stretchable, freezing-tolerant ionic conductive organohydrogel for multi-functional sensors. Advanced Functional Materials. 2020;**30**:2003430

[53] Liang Y, Ye L, Sun X, Lv Q, Liang H. Tough and stretchable dual ionically cross-linked hydrogel with high conductivity and fast recovery property for high-performance flexible sensors. ACS Applied Materials & Interfaces. 2020;**12**:1577-1587. DOI: 10.1021/acsami.9b18796

[54] Jia Z, Zeng Y, Tang P, Gan D, Xing W, Hou Y, et al. Conductive, tough, transparent, and self-healing hydrogels based on catechol–metal ion dual self-catalysis. Chemistry of Materials. 2019;**31**:5625-5632. DOI: 10.1021/acs.chemmater.9b01498

[55] Devaki SJ, Narayanan RK, Sarojam S. Electrically conducting silver nanoparticle–polyacrylic acid hydrogel by in situ reduction and polymerization approach. Materials Letters. 2014;**116**:135-138. DOI: 10.1016/j.matlet.2013.10.110

[56] Zhou Y, Wan C, Yang Y, Yang H, Wang S, Dai Z, et al. Highly stretchable, elastic, and ionic conductive hydrogel for artificial soft electronics. Advanced Functional Materials. 2019;**29**:1806220. DOI: 10.1002/adfm.201806220

[57] Keplinger C, Sun J-Y, Foo CC, Rothemund P, Whitesides GM, Suo Z. Stretchable, transparent, ionic conductors. Science. 2013;**341**:984-987. DOI: 10.1126/science.1240228

[58] Nagamine K, Chihara S, Kai H, Kaji H, Nishizawa M. Totally shape-conformable electrode/hydrogel composite for on-skin electrophysiological measurements. Sensors and Actuators B: Chemical. 2016;**237**:49-53. DOI: 10.1016/j.snb.2016.06.076

[59] Sun J-Y, Keplinger C, Whitesides GM, Suo Z. Ionic skin. Advanced Materials. 2014;**26**:7608-7614. DOI: 10.1002/adma.201403441

[60] Pan L, Cai P, Mei L, Cheng Y, Zeng Y, Wang M, et al. A compliant ionic adhesive electrode with ultralow bioelectronic impedance. Advanced Materials. 2020;**32**:2003723

[61] Wu H, Yang G, Zhu K, Liu S, Guo W, Jiang Z, et al. Materials, devices, and systems of on-skin electrodes

for electrophysiological monitoring and human–machine interfaces. Advancement of Science. 2021;**8**:2001938

[62] Chen L, Wang Z, Zhan Z, Xie M, Duan G, Cheng P, et al. 3D printed super-anti-freezing self-adhesive human-machine interface. Materials Today Physics. 2021;**19**:100404

[63] Zhang Q, Liu X, Duan L, Gao G. Nucleotide-driven skin-attachable hydrogels toward visual human–machine interfaces. Journal of Materials Chemistry A. 2020;**8**:4515-4523

[64] Zhang D, Qiao H, Fan W, Zhang K, Xia Y, Sui K. Self-powered ionic sensors overcoming the limitation of ionic conductors as wearable sensing devices. Materials Today Physics. 2020;**15**:100246

[65] Wu Z, Yang X, Wu J. Conductive hydrogel-and organohydrogel-based stretchable sensors. ACS Applied Materials & Interfaces. 2021;**13**:2128-2144

[66] Song J, Chen S, Sun L, Guo Y, Zhang L, Wang S, et al. Mechanically and electronically robust transparent organohydrogel fibers. Advanced Materials. 2020;**32**:1906994

[67] Chen F, Zhou D, Wang J, Li T, Zhou X, Gan T, et al. Rational fabrication of anti-freezing, non-drying tough organohydrogels by one-pot solvent displacement, Angew. Chem. 2018;**130**:6678-6681

[68] Wu J, Wu Z, Xu H, Wu Q, Liu C, Yang B-R, et al. An intrinsically stretchable humidity sensor based on anti-drying, self-healing and transparent organohydrogels. Materials Horizons. 2019;**6**:595-603

[69] Shi S, Peng X, Liu T, Chen Y-N, He C, Wang H. Facile preparation of hydrogen-bonded supramolecular polyvinyl alcohol-glycerol gels with excellent thermoplasticity and mechanical properties. Polymer (Guildf). 2017;**111**:168-176

[70] Gao H, Zhao Z, Cai Y, Zhou J, Hua W, Chen L, et al. Adaptive and freeze-tolerant heteronetwork organohydrogels with enhanced mechanical stability over a wide temperature range. Nature Communications. 2017;**8**:1-8

[71] Xie W, Duan J, Wang H, Li J, Liu R, Yu B, et al. Ultra-stretchable, bio-inspired ionic skins that work stably in various harsh environments. Journal of Materials Chemistry A. 2018;**6**:24114-24119

[72] Liu R, Zhao S, Liu J. From lithographically patternable to genetically patternable electronic materials for miniaturized, scalable, and soft implantable bioelectronics to interface with nervous and cardiac systems. ACS Applied Electronic Materials. 2021;**3**:101-118. DOI: 10.1021/acsaelm.0c00753

[73] Green R, Abidian MR. Conducting polymers for neural prosthetic and neural interface applications. Advanced Materials. 2015;**27**:7620-7637

[74] Kotov NA, Winter JO, Clements IP, Jan E, Timko BP, Campidelli S, et al. Nanomaterials for neural interfaces. Advanced Materials. 2009;**21**:3970-4004

[75] Oh W-K, Kwon OS, Jang J. Conducting polymer nanomaterials for biomedical applications: cellular interfacing and biosensing. Polymer Reviews. 2013;**53**:407-442

[76] Che J, Xiao Y, Zhu X, Sun X. Electro-synthesized PEDOT/glutamate chemically modified electrode: a combination of electrical and

biocompatible features. Polymer International. 2008;**57**:750-755

[77] Guimard NK, Gomez N, Schmidt CE. Conducting polymers in biomedical engineering. Progress in Polymer Science. 2007;**32**:876-921

[78] Harris AR, Morgan SJ, Chen J, Kapsa RMI, Wallace GG, Paolini AG. Conducting polymer coated neural recording electrodes. Journal of Neural Engineering. 2012;**10**:16004

[79] Guzzo S, Carli S, Pavan B, Lunghi A, Murgia M, Bianchi M. Evaluation of the in vitro biocompatibility of PEDOT: nafion coatings. Nanomaterials. 2021;**11**:2022

[80] Ouyang J. Application of intrinsically conducting polymers in flexible electronics. SmartMat. 2021;**2**:263-285

[81] Cuttaz EA, Chapman CAR, Syed O, Goding JA, Green RA. Stretchable, fully polymeric electrode arrays for peripheral nerve stimulation. Advancement of Science. 2021;**8**:2004033

[82] Abidian MR, Martin DC. Experimental and theoretical characterization of implantable neural microelectrodes modified with conducting polymer nanotubes. Biomaterials. 2008;**29**:1273-1283

[83] Saunier V, Flahaut E, Blatché M-C, Bergaud C, Maziz A. Carbon nanofiber-PEDOT composite films as novel microelectrode for neural interfaces and biosensing. Biosensors & Bioelectronics. 2020;**165**:112413

[84] Liu X, Liu J, Lin S, Zhao X. Hydrogel machines. Materials Today. 2020;**36**:102-124

[85] Goding J, Vallejo-Giraldo C, Syed O, Green R. Considerations for hydrogel applications to neural bioelectronics. Journal of Materials Chemistry B. 2019;**7**:1625-1636

[86] Chung J, Khot A, Savoie BM, Boudouris BW. 100th anniversary of macromolecular science viewpoint: Recent advances and opportunities for mixed ion and charge conducting polymers. ACS Macro Letters. 2020;**9**:646-655